Testing and Measurement

A Practical Guide for Classroom Teachers and Aspiring Educational Leaders

First Edition

Melanie A. DiLoreto

cognella®
SAN DIEGO

Bassim Hamadeh, CEO and Publisher
John Remington, Executive Editor
Gem Rabanera, Senior Project Editor
Celeste Paed, Associate Production Editor
Jess Estrella, Senior Graphic Designer
Trey Soto, Licensing Coordinator
Natalie Piccotti, Director of Marketing
Kassie Graves, Vice President of Editorial
Jamie Giganti, Director of Academic Publishing

Printed in the United States of America.

cognella® ACADEMIC PUBLISHING
3970 Sorrento Valley Blvd., Ste. 500, San Diego, CA 92121

Table of Contents

Preface

In the United States, the method of assessing what students know and can do has not significantly changed in more than 60 years. To shift toward an approach to assessment that focuses on outcomes-based student learning, there is a need for explicit and transparent processes shared among educators and students. Assessment is a key factor in decision making. It is imperative that decisions are made based on the results of quality instruments, transparency, and a shared vision of the purpose of assessment.

I hope this book will help professional educators begin the dialogue about the purpose of assessment, its link to student learning, and its impact on decision making. As you read more about issues related to assessment, testing, measurement, and to the accountability movement, it is vital that professional educators realize the importance of testing and the impact of the decisions made based on those tests. This book will provide multiple tools that may be utilized to determine the quality of your own assessments as well as commercially developed tests.

CHAPTER 1

Assessment and Accountability in Education

Does it really matter? A historical perspective

AS YOU READ

As you read the chapter, think about the following:

- How do **you** define or explain assessment and accountability?
- What is **your** experience with assessment and accountability?

WHAT ARE THE GENERAL PRINCIPLES OF ASSESSMENT AND ACCOUNTABILITY?

Educational Standards Movement

Educational reform is an ongoing topic among policymakers. Over the years, the emphasis of national standards has shifted from inputs to processes to outcomes. Indeed, a noticeable shift from inputs to outcomes has occurred over the past 2 decades as can be seen in the requirements established by policymakers. Due to this shift, schools began to demonstrate that their teachers are not only qualified, but that they also use research-based best practices as well as demonstrated that their students attained the learning outcomes. Thus, national standards reform efforts focus on evaluation and accountability of schools with a current emphasis on outputs or outcomes of education rather than the inputs. This shift is evident in education reform from the late 1990s into the early 2000s wherein the focus turned to student learning outcomes. Specifically, Goals 2000, a key education initiative of the Clinton administration, encouraged states to develop content and performance standards that were demanding, shifting the focus to outcomes of education.

Educational policymakers consistently focus discussions and legislative mandates on institutional effectiveness defined as "the systematic, explicit, and documented process of measuring performance against mission in all aspects of an institution" (SACS, 2005; SACSCOC, 2012, p. 115). Typically, these policies are meant to be used as a way to encourage accountability. One measure often chosen by policymakers to determine the effectiveness of a school is how well its students perform on various assessments. Therefore, as the educational landscape continues to shift from inputs to outcomes, the need for improved performance on assessments becomes more evident. High-stakes assessment results seem to have become the key measure of the outcomes in today's educational climate. Consequently, with the growing emphasis on accountability, it is as important as ever to ensure that students not only know the content but also are able to effectively perform in jobs post-graduation. Thus, assessment of students performed and designed by educators is an important component in the shifting accountability landscape.

K–12 Standards-Based Curriculum Reform

Curriculum standards have become a formidable force affecting the reformation of education at all levels since the publication of the document *A Nation at Risk* (National Commission on Excellence in Education [NCEE], 1983) and even more recently with the adoption of the Common Core Standards Initiative (Common Core State Standards Initiative [CCSSI], 2020). These curriculum standards were written with an emphasis on experiences (processes) and outputs (outcomes). Using these curriculum standards as a major measure, educators began to adopt and use research-based teaching methods instigating a trend toward a hands-on constructivist approach to student learning, wherein the learning process is emphasized. This emphasis has had a major impact on current teachers because they are required to be facilitators of the learning process rather than transmitters of knowledge.

Although many educators agree that standards-based education is the central driving force in educational reform today, there remains much debate on the meaning of standards-based education. For the most part, educators agree that content standards are subject-matter descriptions of what students should know or be able to do within specific grade levels; however, these are often confused with performance standards that are typically interpreted as expected performance on a test. Although many educational policymakers emphasize systematic reform in schools, it is unclear how standards-based reforms are expected to work. As a result, many classroom teachers and school administrators use the standards as their curriculum instead of guiding the curriculum. Textbooks often drive what is taught in classrooms instead of teachers determining what is best for their students based on variables proven to influence student learning. To further complicate matters, policymakers and other stakeholders are calling for higher achievement and effectiveness at all levels, both in teaching and learning; however, it is unclear how to best assess these. Consequently, the reality of the implementation of standards-based education has resulted in what is often considered the cheapest and easiest approach—standardized test-based accountability.

Often, state and federal policymakers implement educational reform hoping that these reforms will actually improve students' learning and ultimately their academic achievement. Thus, teachers and leaders today must not only be prepared to teach and lead their students, but also be prepared to be held accountable for the student learning outcomes resulting from their teaching and leadership methods.

Accountability in Education

Accountability is not a new concept within the United States. The root of accountability movement can be traced back to the Soviet Union's launch of Sputnik in 1957. In 1958, almost a year after the Soviets launched this satellite, the National Defense Education Act (NDEA) was passed under President Dwight Eisenhower's administration. This act provided increased federal funding for education, especially in the areas of mathematics and science. Barely 7 years later, Congress passed another piece of legislation that included increased funding as well as requirements of accountability. This new legislation, titled the Elementary and Secondary Education Act (ESEA), was passed by Congress in 1965 under President Lyndon Johnson's administration and was deemed to be the first shot in what has been termed the *war on poverty* as outlined by President Johnson in his January 1964 State of the Union Address.

The ESEA, the largest federal funding allocation to date, was primarily influenced by the disparity in educational opportunities and in student performance. After the passing of such a federal mandate, legislators demanded accountability for the funds distributed to various programs. As a result, there was an enormous increase in the number of norm-referenced tests developed and published to determine student learning as a means to make the educational system accountable. In addition to the increased use of standardized tests, the Normal Curve Equivalent (NCE) was developed to standardize scores to coincide with national percentile ranks at three points (1, 50, and 99). Furthermore, during the 1970s several state legislative bodies enacted minimum-competency testing requirements that were intended to ensure state-mandated standards were being met. Indeed, the genesis of these minimum-competency tests was shortly after the initial implementation of norm-referenced standardized testing in the early 1970s.

In 1983, standardized testing in K–12 settings became the emphasis on the national forefront due to the publication of *A Nation at Risk*. By the mid-1980s, every state in the nation had imposed some form of legislative mandate that required accountability measures in K–12 education. In the late 1980s and the early 1990s, various stakeholders, including policymakers and educators, used these standardized test results for accountability purposes. While taking on a variety of forms, they all shared in the aspect that there were increased stakes for educational administrators, instructors, and students in K–12 educational settings.

External accountability pressures make it even more necessary to work toward a system that can be used in the complex and diverse system of education found in the United States. Often, a gap exists between policy and practice in many schools. Information published by the National Center for Higher Education Management Systems (NCHEMS, 2008), indicates that for every 100 ninth-graders, it is estimated that 70 of them will graduate from high school on time, 44 of them will immediately enroll in college, 30 will still be enrolled during their sophomore year, and 21 of them will graduate from college on time. Indeed, the implementation of such a system of accountability measures might increase student performance and concurrently increase the on-time graduation rate.

Differences exist in how and what accountability measures should be established. For example, some believe that legislators have no right to implement policies affecting accountability, but that educators do have such a right because they have the understanding of teaching and learning. Often, teachers are angry and afraid due to the underhandedness of policymakers instituting policies affecting education without the presence of the educators and that teachers believe there is a disregard of their professional expertise. These same teachers, however, believe that they can use their knowledge and expertise to retain their own interpretations of what is best for the organization.

Transparency is a key ingredient for success with assessment and accountability (Jonsson, 2014; Jonsson & Prins, 2019). Clearly, one of the primary goals of assessment is to positively impact student learning. But, to impact student learning positively, it is imperative that all stakeholders share a belief in attainment of this goal. Consequently, for educators and students to meet accountability goals such as positively impacting student learning, individual stakeholder's beliefs about assessment should be acknowledged. Furthermore, interpretation of the purpose of assessment becomes problematic when there is a disconnect between the initiatives of policymakers and the conceptions held by educators and students alike. Simply implementing assessments does not necessarily improve student learning. Coming to an understanding of the purposes of assessment make it possible to establish an accountability system that can potentially increase student learning.

Beliefs influence both behavior and motivation. As such, it may be necessary to consider how social expectations of peers and cultural transmission may be effective ways to transform beliefs. There is research that indicates beliefs are only changed when they are no longer satisfactory to an individual (Pajaras, 1992). Beliefs become unsatisfactory only when they are challenged to the point where they cannot be assimilated. Changes in beliefs are possible when supported by the culture of an organization. Thus, because of their distinct roles and their knowledge, classroom teachers and educational administrators inherit a responsibility to respond to policy initiatives in a way that takes into account the current beliefs held by both educators and students. Allowing time for teachers and administrators to discuss implications of new policy initiatives increases the likelihood of a change in beliefs of these educators. Furthermore, a policy's message can be thought of as an external representation that demonstrates that problems have been targeted, delineated, and well defined. Inconsistencies and changes in state policies can create a sense of anxiety that can then influence the culture of an organization. Thus, organizational learning fostered by transparent discussions may be a critical component to embracing a culture of assessment to improve student learning.

The importance of meeting the criteria suggested above becomes evident when in an era of increased accountability, policymakers often use assessment data to determine overall effectiveness as well as specific student learning outcomes. It is the case that accountability measures exist for classroom teachers, educational leaders, and students and that there are often high-stakes decisions made as a result of these accountability measures. For example, classroom teachers are expected to demonstrate strong, research-based instructional skills. Their annual evaluations are often based on a statistical formula undergirded by their students' results on standardized tests. Decisions about students' progression through schooling and ultimately graduation are made based on individual student performance. Sometimes, however, these high-stakes decisions are contradictory to the intended outcomes of initial accountability mandates and accompanying assessments. Often, the use of high-stakes accountability confounds the intended positive effects of accountability.

To meet the demands of accountability, policymakers have focused on the use of high-stakes, standardized tests and assessments for nearly three decades. It is believed that the reason for this reliance on standardized tests and assessments is because they are relatively inexpensive compared with other reforms, can be externally mandated, can be rapidly implemented, and the results are fairly immediate and visible. As such, statewide standardized assessments are often used in K–12 settings as a means to determine grade progression as well as graduation. As a result, tests and assessments foci remain at the forefront of the educational landscape. Mandates such as No Child Left Behind (NCLB, 2002), ESEA (1965), and to a lesser extent, Every Student Succeeds Act (ESSA, 2015) often require K–12 school systems to provide evidence of meeting the policy requirements by providing various data and assessment results.

High-Stakes Assessment in K–12: What Does Assessment Really Mean?

During the past 50 years, the accountability movement has spawned an ever-growing focus on assessment. James B. Conant, past president of Harvard University who was known for his egalitarian vision of education, provided a rationale for differentiated instructional programs based on a common core in the 1950s. By 1986, all 50 states were under some form of a legislative mandate requiring assessment of K–12 students. Although there is no one definition of the term assessment and there is no one single purpose of assessment, educators often use assessments to make decisions about student knowledge. Various researchers have developed their own meanings of assessment. Some refer to assessment as the data collected by specific tools and measures that are used to meet external accountability requirements established by various accrediting agencies. Others contend that assessment consists of a composition of both quantitative and qualitative data as well as the interpretation of those data. Yet others recommend that assessment be comprehensive with an iterative cycle of measuring progress at multiple points in time. Some recommend that assessment should focus primarily on student learning and the academically related activities that students engage in during their tenure at an academic institution, specifically analyzing a graduates' workforce and general education skills, domain-specific knowledge, soft skills such as teamwork and creativity, and student engagement.

The National Council of Teachers of Mathematics (NCTM, 1995; 2005) added to the dialogue of recommended uses of assessment by stating that the primary purpose of any student assessment should be to improve student learning. But, the organization contends that there are several secondary purposes for assessment. Specifically, these secondary purposes include: to provide individual feedback to students about their learning; to provide information to the instructor about how well the class understands a particular topic, what additional activities might need to be introduced, or whether it is time to move on to another topic; to provide diagnostic information to instructors about individual students' understanding or difficulties with understanding new material; to provide information to teachers about students' perceptions and reactions to the class, the material, the subject matter, or particular activities; to provide an overall indicator of students' success in achieving course goals; and to help students determine their overall strengths and weaknesses in learning the course material. It is a dilemma that the term assessment is often used within different contexts and with different meanings.

Making the Connection: Policy and Practice

For policymakers, legislators, or educators to develop a clear, simple system of assessment that includes a common language, a common purpose of assessment must be developed. Educational policymakers in the United States over the past several decades have implemented many federal and state mandates requiring more accountability. Assessments are often used to make high-stakes decisions regarding both teachers and students. Although the accountability and assessment movement has been at the forefront of the educational reform landscape, there has been little movement from educators toward embracing the change and simply introducing or mandating a new assessment policy will not necessarily achieve the intended policy objectives. Teachers' attitudes toward and expertise in assessment impacts the way they implement their assessments in the classroom. Additionally, students' attitudes toward and their experiences with assessment affect their personal approach to learning and their beliefs toward future successes as a learner. It is imperative that the development and implementation of an assessment system relies heavily on a clear purpose of assessment and that all stakeholders be included in the design of any system used to make high-stakes decisions.

WHAT DOES THE LITERATURE SAY ABOUT ASSESSMENT AND ACCOUNTABILITY?

Reading 1.1

INTELLIGENT ACCOUNTABILITY

Re-Thinking the Concept of "Accountability" in the Popular Discourse of Education Policy

Scott Ellison

> I believe that the quality of our education system says as much about the long-term health of our economy as the stock market, the unemployment rate and the size of the gross domestic product. That's because the quality of our work force and the intellectual breadth and depth of future leaders is directly related to the quality of education we provide today. So I begin ... *by recognizing America's common agenda to promote economic security through education.* (U.S. Secretary of Education Arne Duncan)

In prepared remarks before the United States Chamber of Commerce's "Education and Workforce Summit" in November of 2009, U.S. Education Secretary Arne Duncan expressed a sentiment now dominant in the popular discourse of education reform in the United States linking the economic fortunes of the U.S.A. in the global economy to the human capital produced by its education sector.[1] This perspective envisions economic success as emerging from an educational system that produces new generations of technological and scientific innovators and high-skill workers to drive an economic sector geared toward providing high value-added goods and services on a global marketplace.[2] For K–12 schooling, the challenge would appear to be to produce students possessing the foundational knowledge and skills of literacy, mathematics, and critical thinking required for university and technical training.

To that end, Duncan offers straightforward proposals for educational reform that require states to set rigorous, easily-measured academic standards, create assessments linked to those standards, use data generated by those assessments to raise academic achievement, and, finally, to hold educational stakeholders [administrators, teachers, and students] accountable for failure. Most commonly subsumed under the conceptual rubric of 'accountability,' this model of education reform has become the default policy position of both political parties in the U.S. and has become the commonsensical position in popular debates over education reform. It is a policy position built on the assumption that it is possible to dramatically raise academic achievement in public education by re-organizing its incentive structures. However, critics of these reform policies charge that placing an undue emphasis on easily defined academic standards tied to standardized assessments and accountability regimes will lead to a narrowing of school curriculum to the easily assessed, the regimentation of schooling, the re-enforcement of teacher-centric pedagogies that provide little opportunity for student discovery and critical thinking, and an over-reliance on rote learning. It is from this indeterminate situation in debates over education policy that this inquiry finds its impetus.

The task of this article is to unpack the concept of accountability in order to clarify and critique the logic of this educational and political concept. To accomplish this task, I will employ a synthetic method of analysis that will, first, situate accountability within the larger framework of standards-based education reforms of which it is an integral element. From there, the second step is to examine the research literature in order to interrogate standards-based policy reforms at each point in its logical chain so as to unpack the unquestioned assumptions and problematics inherent to the concept of accountability that are often obscured by contemporary educational discourse. The results of this study suggest that critics of accountability policies are well justified in their concerns.

A SYNTHETIC METHOD OF INQUIRY

The methodology for this inquiry is grounded in an ontological observation into the internal contradictions between the conceptual norms with which modern societies take their general orientations and the concrete realities generated by those conceptual understandings. This methodology is itself rooted in a long tradition of praxis philosophy tracing its lineage to the works of Hegel.[3] In its most general sense, the synthetic mode of inquiry first outlined by Hegel and employed in this essay takes as its subject the concepts, theories, and ideas that represent the 'spirit of the age,' the conceptual knowledge through which individuals become conscious of their worlds and from which societies seek justification for concerted social action and governmental policy.

A synthetic method of inquiry can be characterized as being what Dewey called a "double movement" to and from meaning in which critique is located within the reflexive movement between the conceptual and concrete.[4] Synthesis begins with a practical concept at work in popular discourse as it is most readily present as a discursive practice. From there the inquirer pulls apart the concept through an analysis of the concrete realities that is its subject. The product of this deductive movement establishes the foundation for the transformation of the concept under analysis, and hence the concrete realities appropriate to it, by providing the inquirer with the raw material for an inductive movement of re-conceptualization.

The synthetic mode of inquiry employed in this essay denotes a heuristic consisting of three "moments" of inquiry (definition, division, and synthesis) in which the inquirer works through the conceptual logic of the subject of inquiry in order to unlock its transformative potential, to put the subject of inquiry "to the test of reality, of contemporary reality, both to grasp the points where change is possible and desirable, and to determine the precise form this change should take."[5] The "double movement" of a synthetic analysis begins with an indeterminate situation in which cemented "common sense" ideas run aground the concrete realities of a contradictory, dynamic present-history.

> Moving from an indeterminate situation of conceptual failure and contradiction, inquiry begins with a definition of a concept in its most immediate form. From definition, inquiry moves through the logic of this conceptual framework in a deductive movement that articulates the complex determinants and processes conditioning its actualization in the concrete [division]. Moving inductively from this empirical base, the concept is then synthesized into a conceptual framework through which the indeterminate situation from which inquiry began is resolved and through which the epistemic validity of its normative claims are to be established.[6]

Thus, a synthetic mode of inquiry returns to its starting point as a rich totality of determinations built around a rigorous investigation of the empirical evidence relevant to the subject of inquiry and the reconciliation of the conceptual norms implicit to the subject of inquiry and the practices that seek justification in those normative ideals.

In this inquiry, I will begin by first defining the concept of accountability by situating it within the standards-based education reforms of which it is an integral element. To do so, I will examine popular texts produced by prominent advocates of standards-based reforms so as to flesh out the concept of accountability as it is employed in popular debates over education policy and reform. From there, I will peer beneath the veneer of standards-based education reforms to examine the efficacy of those reforms in achieving the academic gains and functional rationality from which it seeks justification. I will accomplish this by examining peer-reviewed research literature so as to test the epistemic validity of accountability at each point in its logical chain, identifying moments of contradiction and epistemic failure. From there, I will then work back through the conceptual logic of accountability in order to establish the foundation for the reconciliation between the normative ideals animating the concept of accountability and the practices appropriate to those ideals. In the final section, I will conclude the article with a critical reflection that seeks to flesh out some of the larger lessons to be taken away from this synthetic analysis.

STANDARDS, ASSESSMENT, AND ACCOUNTABILITY

Returning to Duncan's remarks, it would appear that the key to ensuring that students are taught the foundational knowledge and skills of literacy, mathematics and critical thinking required for success in a 21st century economy begins with a set of rigorous academic standards for those subjects. Kosar makes the case for a standards-based approach to reforming educational practices this way:

> [T]here is a powerful [behaviorist] logic to standards-based reform: children will not learn to high levels unless they are taught challenging curricula ... [T]o raise achievement, the level of skills and knowledge students are taught must be raised, and this can be done through establishing challenging education standards. Doing this will maximize the probability of good teaching or worthwhile content to all students. And the children will respond.[7]

The working assumption of this approach is that creating clearly defined, rigorous academic standards aligned to high-stakes assessments will push schools toward sound curricular and pedagogical practices that will then lead to higher student achievement. If the nation's system of public education is to meet the challenges of globalization, then federal and state policy makers should set high standards for achievement to ensure that students are taught the foundational knowledge they will need to compete in an information-based economy.

To ensure that students, administrators, and schools maintain this high level of achievement, a 21st century system of public education requires a rigorous regime of performance-based accountability. Such a regime begins with the necessary "mission, values and guiding principles" required to create a broad framework of academic and performance standards for students, administrators, and schools.[8] These standards would form the foundation for the design of instruction, curricula, and high-stakes assessments[9] "that have real consequences for professionals in the schools."[10] Hess observes: "Performance information is meaningful only when it is backed by real incentives; when principles and superintendents use it to reward or penalize teachers based on how well they are serving their students."[11] For teachers, accountability means that their performance evaluations would be linked to student performance on standardized assessments. Likewise, administrators and entire schools would be judged by student performance on standardized assessments with the explicit threat of dismissal of administration or the possibility of a complete re-structuring of schools either by quasi-privatization (e.g., charter schools) or re-constitution of school staff.

BENEATH THE VENEER: RESEARCH LITERATURE

Beneath the veneer of these standards-based reforms lies an assumption of a functional reflexivity. Clear, measurable academic standards provide the foundation for the construction of rigorous curricula, sound instructional models, and assessments that will challenge students to achieve at high levels. Standardized assessments measure students' achievement in reaching the prescribed academic standards, measure teacher and school effectiveness, and provide the necessary data for realigning curricular and pedagogical practices in order to address students' academic needs. In turn, the tough accountability measures attached to performance on those assessments provide incentives for students, teachers, and administrators alike in ensuring student achievement. Put more simply, the assumption here is that standards-based reform will foster a rational organization of schooling that will increase student achievement. It is an assumption of reinforcing incentives where standards are the carrot accompanying the accountability stick and rational education decisions are made possible by the empirical data provided by objective assessments. However, a review of the research literature offers little to support these assumptions.

Sloan and Kelly identify two controversial issues involved with standards-based reforms and high-stakes assessments.[12] The first of these issues lie with the desirability [or even possibility] of developing clear, easily measured standards as the foundation of instruction, assessment, and accountability.

> Content mastery at some level is a cognitive event: the understanding of powerful, complex, and sometimes fuzzy ideas. For that reason, at least for challenging content, it may be difficult to write clear and simple standards, thereby making their operationalization for curriculum development, test construction (of any genre, objective or authentic), and alignment between the two problematic.[13]

On the one hand, the logic of a standards-based framework requires challenging academic content as an incentive for high academic achievement; on the other hand, standardized assessments require clear, easily measured criteria by which to judge student success. Indeed, *prima facie*, it stands to reason that in the process of articulating challenging academic content into clear, easily-measured standards a good deal of the complexity, ambiguity, and contradictions that make content challenging to begin with would be lost. More importantly, even assuming the desirability of constructing clear, easily-measured academic standards, the types of assessment most often associated with standards-based reforms raise issues that call into question the assumption of reflexivity between these assessments and educational decision-making processes.

Sloan and Kelly point out that different types of assessments test different skills and serve different purposes. The multiple choice assessments associated with standards-based reforms in the U.S. are most often constructed within a pyschometric framework designed to produce "economically tractable and defensible reliability indices" for ranking and norming purposes.[14] Constructed along a Bell curve, it would appear that these types of assessments are far better suited to scaling students than providing useful information capable of improving student learning. Yet a key assumption of standards-based reform is that these "scientifically" constructed assessments must do exactly that, i.e., improve student learning.

At the district and state level, standardized assessments are being increasingly used to differentiate between effective and ineffective schools and individual teachers. While this may seem a rather straightforward and intuitive means of measuring effectiveness, it is far from rigorous. Measuring school proficiency/deficiency based on standardized assessments assumes a causal relationship between differences in student test performance and differences in the effectiveness of instruction. A school accountability framework based on standardized test scores alone fails to recognize other academic inputs, such as student-teacher characteristics, instructional practices, resources, curriculum, etc., involved in academic achievement and test performance. Robert Linn notes that, despite the rhetoric of many reformers, using standardized test scores to distinguish between effective and ineffective schools is decidedly unscientific.[15] As a tool for judging the effectiveness of individual teachers, the National Academy of Sciences warns against the over-emphasis of standardized assessment scores as the primary mode of evaluation in Duncan's *Race to the Top Initiative*, noting that, "while standardized tests are helpful in measuring a reform's effects, evaluations should rely on multiple indicators of what students know and can do, not just a single test score."[16] Further, in a more general sense, determining proficiency and deficiency using data from standardized assessments must *a prori* involve some sort of ranking process that requires a determination of "cut points" in the distribution of student scores that is, as Andrew Ho notes, necessarily "judgmental."[17]

Rupp et al. examine the common approach of standardized assessment designed to measure reading comprehension skills, which is to ask students to respond to a text passage with multiple-choice questions. Their findings suggest that students approach these questions as problem solving tasks and that these assessments actually test a rather limited range of reading skills. "[D]ifferent [multiple choice] questions do not merely tap but, indeed, create very particular comprehension and response processes. Therefore, a blanket statement such as '[multiple choice] questions assess reading comprehension' is nonsensical for any test."[18]

In order to evaluate the efficacy of standardized assessments as pedagogical tools Rupp and Lesaux compared measures of reading achievement produced by standardized assessments and the component skills of literacy as identified by a "diagnostic battery of standardized and experimental tasks."[19] Their findings suggest that the proficiency classifications of standardized assessments mask a great deal of heterogeneity of reading skills at all levels. This raises two serious issues. First, it questions whether standardized assessments provide adequate measures of student reading comprehension skills, and, second, it calls into question whether standardized assessments provide actionable information for improving student achievement. "Given the reliance on standards-based assessments to guide educational decision-making ... there is a need to seriously consider whether the properties of these tests support any interpretation at the level of the individual and similarly whether there is any instructional information to be gleaned from the results."[20]

Looking at the component reading comprehension skills of students identified as failing by standardized assessments, Bully and Valencia also find that standardized assessments mask variations in the component skills of reading.[21] By masking the considerable heterogeneity of student reading skills, standardized assessments appear not only to provide a limited picture of students' reading comprehension, but they also fail to provide the necessary data for tailoring instruction to students' specific needs and contribute to student achievement. Further, Bully and Valencia question the utility of policy decisions mandating specific instructional strategies and curricular programs for failing schools, such as Reading First, that are becoming increasingly common in standards-based reforms. "Our data suggest that such policies aimed at specific, predetermined, instructional approaches cannot possibly fit the various needs of at-risk children."[22] In short, on both the student and school levels, it appears that standardized assessments are being put to tasks for which they are neither designed nor well-suited.

Turning now to the assumption that the incentive structure created by attaching high-stakes consequences to standardized assessments will lead to achievement gains, the research literature offers little supporting evidence. Examining state-level data, Nichols et al. investigate the relationship between high-stakes assessments and student achievement. They find that assessment pressures have a positive impact on 4th grade mathematics but no impact on 4th grade reading or 8th grade reading and math, suggesting that accountability pressure may only impact skills that can be effectively taught in a direct instructional approach.

> [O]ur findings (and lack of findings) lead us to the conclusion that high-stakes testing pressure might produce effects only at the simplest level of the school curriculum: Primary school arithmetic where achievement is most susceptible to being increased by drill and practice and teaching to the test.[23]

Similarly, Merchant et al. find limited evidence that high-stakes assessment policies raise achievement only in the most basic skills but no evidence that tough accountability policies raise achievement in more advanced subject areas.[24] Lee and Wang find no evidence of achievement gains associated with high-stakes assessments nor a significant change in the distribution of achievement across student groupings.[25] Overall, results from my review of research literature examining state-level data do not support "any argument that high-stakes testing is necessary to raise student achievement."[26]

Interestingly, there is a growing number of studies that suggest that not only do high-stakes assessments have little impact on gains in student achievement overall but that implementation of these policies leads to strategic responses by educational players that generate significant social costs. Linking rigorous accountability regimes to standardized assessments creates powerful incentives to narrow instruction to only those curricular areas being assessed and to the wide-spread adoption of teacher-centric 'drill and kill' instruction.[27] More troubling, there is evidence that the incentives created by standards-based accountability reforms lead to perverse outcomes and gaming strategies. Through grade retention and inadequate reporting of graduation rates, high schools serving even the most challenging student bodies were able to raise average student achievement, but it is a success that appears to come at the cost of the lowest performing students, who eventually disappear or drop-out.[28]

In sum, the research literature fails to provide compelling evidence that the high-stakes tests and tough accountability measures associated with standards-based reform increase student achievement. Even assuming the desirability of constructing clear, easily measured standards, the assessments aligned to those standards fail to provide the rich data educators require to raise the achievement of struggling students. More troubling, the incentives created by the accountability measures associated with standards-based reforms appear to generate significant social costs in the form of higher retention rates and increases in the number of drop-outs. Thus, the functional reflexivity assumed by the standards-based reforms is fractured at each point in its logical chain. The concept of accountability appears to be internally contradictory and flawed. However, at each point of conceptual failure, the division of accountability demonstrates not only the failure of standards-based reforms to fulfill their own normative goals. It also provides us with a road map for constructing new policy ideas capable of making those goals an empirical reality.

SYNTHESIS

If we take the mathematical-scientific knowledge, literacy, and creative-innovative thinking advocated by standards-based reformers as being the foundation for an intelligent system of accountability, then the preceding examination of the popular conceptualization of accountability opens up moments of possibility to break out of cemented conceptions and practices to develop new policy approaches capable of achieving the normative goals of preparing students for the intellectual demands of the 21st century. It is to this task that we now turn.

Rethinking Standards

As Sloan and Kelly point out, complex ideas and knowledge are necessarily fuzzy in that they denote "cognitive events" as opposed to being bits of data to be stored on organic hard-drives. Mathematical-scientific knowledge denotes more than having a grasp of the mechanics of mathematical computation and the ability to carry out formulaic "experiments" as one might follow a recipe. Instead, it denotes the ability to employ mathematical reasoning and the scientific method in such tasks as developing new technologies, answering scientific questions, and developing new materials. In short, mathematical-scientific knowledge is not an invisible cookbook to be carried around in a cognitive knap sack. It denotes an intellectual skill set that empowers one to utilize mathematical and scientific thinking in productive human activities of problem solving, knowledge construction, and creation.

Likewise, literacy surely denotes more than the simple decoding of information from text or the ability to solve vocabulary and analogy puzzles. Literacy denotes an intellectual skill set. It denotes the ability to interpret and communicate complex ideas through the use of symbolic systems that constitute the core of conscious thought, communication, and intellect. More specifically, literacy involves the decoding, processing, and interpretation of written text in a process of critical thinking in which the outcome is substantially more than the sum of its parts. Literacy involves the ability to organize one's thoughts in a logical framework; to symbolically represent those ideas so that others can decipher them; and to anticipate the needs, perspectives, and skills of an unknowable audience. In short, literacy denotes a mode of conscious thought and symbolic communication.

As a curricular goal, creative-innovative thinking proves to be an even "fuzzier" concept. Denoting more of an intellectual disposition than curricular knowledge in the traditional sense, creative-innovative thinking would appear to demand that the previous two curricular standards, mathematical-scientific knowledge and literacy, be achieved in such a way so as to foster curiosity, a love of learning, problem-solving skills and creative expression. Or, to put it more generally, creative-innovative thinking denotes a mode of intellectual life. Perhaps a better model for this broad curricular goal can be found in Thayer-Bacon's conception of *constructive thinking* as an organizing principle that places equal emphasis on reason, creativity, intuition, collaboration, and care as necessary human tools of creation, innovation, and problem-solving.[29]

Working through the logic of the curricular goals identified by standards-based reform advocates as being the necessary foundation for rigorous accountability through the lens of the research literature allows us to draw some broad conclusions about the nature of the academic standards that are appropriate to the task. Developing standards from the broad curricular goals of mathematical-scientific knowledge, literacy, and creative-innovative thinking would appear to require that those standards be broadly defined and de-coupled from a direct correlation with individual grade levels. If these broad curricular goals denote an intellectual skill set, a mode of conscious thought and symbolic communication, and a mode of intellectual life respectively, then it is clear that standards formulated from those goals must be articulated at a level of generality appropriate to the "cognitive events" they describe. Further, the uneven intellectual development and maturation of individual students would seem to preclude the utility of linking individual standards to any one specific grade level or calendar year. Perhaps a more appropriate model would entail the use of benchmarks situated at transitional points in schooling, such as the transition from primary to secondary schooling. However, what is most important for the current study is to note that the broad curricular goals of creative-innovative thinking, mathematical-scientific knowledge, and literacy do not lend themselves to clearly defined academic standards that can be assessed on large-scale tests employing a multiple choice format. An accountability system constructed on the foundation of these curricular goals must employ an intelligent system of assessment that balances the needs of policy-makers and communities with the learning needs of individual students.

Re-Thinking Assessment

Thus, a constructive system of assessment would deploy a twofold strategy. At the classroom level, student assessment should primarily involve the use of authentic, formative assessments geared toward student learning. The most appropriate form of assessment for this task would be teacher-constructed performative assessments that provide rich data on student knowledge and skills as well as identify the educational needs to be addressed. In order to gear instruction to increased student learning, teachers require an accurate, in-depth perspective on student achievement and academic needs, and this perspective is best acquired through authentic assessments constructed by teachers for individual students according to their needs. These kinds of assessments would provide teachers not only with a rich perspective on student learning but would also provide an opportunity to assess more intangible skills and dispositions such as creativity, curiosity, and attitudes toward learning. For students, formative assessments provide immediate feedback on achievement and identify areas in which they need to focus their studies. Formative assessments provide students with the necessary feedback to develop meta-cognitive understandings of their own skills, knowledge, and interests, and it affords them the opportunity to take ownership of their learning.

At the district and state levels, the goal is to develop rich data on student achievement covering large populations in a manner that is both cost-efficient and practical. A reasonable way forward in achieving this goal is to adopt a balanced approach that employs more authentic assessments to increase the quality of the data available to communities and policy-makers in conjunction with random-sampling techniques that can gather large-scale data while addressing issues of cost and practicality. The quality of the data available to policy-makers and administrators would benefit significantly from the use of more authentic assessments that incorporate elements of interactive, performative, and traditional assessment techniques and are capable of assessing the degree to which students can use knowledge as opposed to their ability to re-produce "dead" content. Utilizing random-sampling techniques will make it possible to carry out fewer assessments and produce high quality data as opposed to current trends toward large-scale assessments that yield relatively poor data. Further, valuable data on student and teacher success can be gleaned from broader measures such as tracking graduation rates, student success rates in the tertiary sector, and parental satisfaction.

Of course, carrying out these reforms would require structural shifts and at least an initial expenditure of resources. At the classroom level, a policy push toward encouraging the use of teacher-constructed performative assessments geared toward advancing student learning would require two actions by policy-makers. First, the wide scale use of authentic, performative assessments is predicated on significant investments in the professionalization of teaching through master-teacher and mentoring programs, professional development opportunities for practicing teachers, and an increased focus in teacher training on constructing and utilizing a variety of assessment techniques to increase student learning. Second, constructing authentic assessments tailor-made to individual students requires that teachers have an in-depth understanding of their individual students' strengths, weaknesses, and academic needs. This would imply that policy-makers should strive to lower student-teacher ratios and experiment with grouping teachers and students together over several academic years, particularly in the primary years. At the district and state levels, implementing large-scale authentic assessments would require a co-ordinated effort to create a system of intelligent standards and appropriate assessments as well as a significant push to increase the number of educational specialists to carry out assessments and interpret the data. Finally, and perhaps most importantly, it is incumbent on policy-makers to provide the necessary resources to make these policy shifts at the classroom, district, and state levels a reality.

Re-Thinking Accountability

Of course, the critical role public education plays in modern society demands that educational actors and institutions be held accountable for fulfilling their social roles and societal mandates. It is both reasonable and desirable that society articulate a uniform set of curricular standards that do justice to the complexity of the cognitive acts they describe; to construct an intelligent assessment regime to evaluate the degree to which individual teachers, schools, and districts are helping students achieve those standards, to reward educational actors for success; and, finally, to hold educational actors accountable for failure. That said, the devil is in the details.

In light of the previous analysis, the questions raised about the efficacy and desirability of accountability policies now dominant in the U.S. requires that researchers and policy-makers remain open to and seek out alternative approaches to education reform that achieve similar goals and ideals as those animating this current era of reform. One alternative approach can be located in one of the most successful education systems in the world in which, interestingly, the concept of accountability is not even a topic of discussion among policy-makers: Finland.[30] The Finnish model of education reform is now widely regarded as having fostered a highly effective, equitable public education system that produces consistent results at a reasonable cost.[31] It is also unique in that, contrary to current international trends toward top-down external accountability measures, the locus of accountability pressures in the Finnish model is centered around the ballot box.

In the midst of an economic crisis in the early 1990's, Finland embarked on an aggressive re-structuring of its educational system from a highly centralized top-down approach toward a system built around a commitment to mutual accountability, professional responsibility, formative assessment, and trust. Finland followed a path of de-centralization coupled with loose national standards that provided teachers and administrators with a high degree of autonomy in achieving those standards and relied on investments in the professionalization of teaching to foster school communities engaged in "research-based teaching" to raise student achievement.[32] In order to balance the needs of policy-makers to provide oversight and teachers to utilize classroom assessment as an instructional tool, Finland embraced teacher-made formative assessments as the dominant method for assessing student achievement and instituted authentic assessments utilizing random-sampling techniques to provide policy-makers with rich data to gauge system and school performance without wide-scale disruption of classroom learning.

Sahlberg notes that the de-centralization of the educational sector and increased autonomy for teachers and school administrators has distributed accountability pressures throughout the system by strengthening a linkage often severed by the test-based accountability pressures now dominating education policy in the United States: professional autonomy and professional accountability.

> Educational reform principles since [the] early 1990s—when much of public sector administration went through decentralization—have relied on building professional responsibilities within schools and encouraging lateral capacity building among teachers and schools, rather than applying external accountability structures. Therefore, sample-based testing, thematic assessments, reflective self-evaluations and putting learning first have established [a] culture of mutual responsibilities and trust.[33]

Despite the fact that the concept of accountability as we think about it in the U.S. is not a topic of discussion among its policy-makers, the Finnish model of education reform offers us a new perspective on how to re-think accountability. To that end, Sahlberg offers three broad recommendations for building what he terms the "intelligent accountability" exemplified by the Finnish model.

- Build trust and shared responsibility by giving school administrators and teachers greater autonomy in developing strategies for achieving academic standards and constructing internal systems of accountability.
- Employ a balanced approach to assessment that provides teachers with the freedom to construct formative assessments as an instructional tool while utilizing authentic assessments [quantitative and qualitative in nature] to provide policy-makers, communities, and parents with rich data to gauge achievement.
- Build lateral capacity by creating spaces within schools for teacher collaboration and research and networking schools to spread ideas throughout the system.

Interestingly, Sahlberg's recommendations for building trust and shared responsibility, a balanced approach to assessment, and lateral capacity also corresponds with the themes that emerged from Darling-Hammond's analysis of successful accountability reforms in the U.S., and it bears a marked resemblance to the conclusions that have emerged from this analysis.[34] While it is impossible to adopt the Finnish reform model wholesale, Finland's educational successes can help educational actors on this side of the Atlantic ocean expand the horizon of the possible in our on-going pursuit of education reform.

In contrast to a punitive system of accountability that relegates school knowledge to the easily assessed, intelligent accountability must work to build the learning capacity of public schooling. If the goal of policy reforms built around the organizing principle of accountability is to ultimately make a positive contribution to student learning and global competitiveness then it would appear necessary to develop an intelligent system of accountability that seeks to build and nurture the institutional capacity of schools. Such a system of intelligent accountability would involve a balanced approach to constructing micro- and macro-accountability structures that provide necessary supports to help teachers and administrators build their professional skills and create productive learning environments for their students. Constructing an intelligent system of accountability would appear to require policy-makers and communities to, at a minimum, carry out the following tasks:

- Coordinate efforts to articulate academic standards from the broad curricular goals of mathematical-scientific knowledge, literacy, and innovative-creative thinking that do justice to the intellectual complexity implicit to each;
- De-centralize the decision-making processes of education policy by providing teachers and school administrators greater professional autonomy in reaching academic standards and constructing internal accountability policies at the school level;
- Make significant investments in teacher training and building lateral capacity across schools, districts, and states;
- Coordinate efforts to construct and implement an intelligent assessment regime that employs a balanced approach of authentic, performative and traditional assessments using random-sampling techniques.

If the overarching goal of public schooling should be the empowerment of students to successfully compete in a new global reality then it is vital that policy-makers and educational leaders make the strategic investments needed to make that a reality. The Finnish model demonstrates that these kind of reforms are both practical and cost effective, and it demonstrates that a de-centralized education system relying on democratic processes to hold education actors accountable for educational outcomes can produce high levels of achievement and equity. However, carrying out such a reform agenda would require an initial expenditure of resources. This again returns us to the role communities and governments should play in an intelligent system of accountability, and it demands critical reflection.

CRITICAL REFLECTION

In contemporary educational discourse, it is now quite common for policy-makers and educational experts to demand that we hold teachers and school administrators accountable for educational failure. However, what is notably absent from those debates is the voicing of a similar need to hold policy-makers and educational experts accountable for providing the necessary resources and know-how to foster academic achievement. There is a great deal of discussion in policy circles and in popular media of how to best hold students and teachers accountable for success and failure, no doubt a tribute to American individualism, but very little is said about holding policy-makers and community leaders accountable for fulfilling their duties in the grand *contrat social* that is public education. In effect, policy-makers and educational leaders have succeeded in pushing accountability "downward" while absolving themselves of being held accountable for fulfilling their vital role in fostering academic achievement in public schools. However, it would be a mistake to view this simply as being a bug in the system. It is not. It is a feature. It is an essential element of the current system.

Education policies which absolve policy-makers and educational experts of accountability for student learning actually work to undermine their own normative justifications. This is not accidental nor is it the inevitable outcome of a structural formation, although that is certainly an element of the equation. The disjunctions and contradictions of current education policies that seek justification in the concept of accountability are a function of a particular articulation of power composed of corporate lobbies, venture philanthropists, think-tanks, conservative political organizations, and advocacy groups that wield educational concepts such as accountability as political tools with frightening efficiency.[35] Indeed, it is the prevalence of these private actors in the social construction of public knowledge and education policy that led me to employ popular texts written by authors connected to national think-tanks currently driving education policy in my initial treatment of the concept of accountability. This articulation of power is a manifestation of a specific form of politics that seeks to subject the public domain to private interests in the pursuit of profit, whether that be through direct per-child payments for schooling "services" or by grabbing a "4 or 5 percent slice of a couple trillion dollar public education real estate market" for speculative investment.[36] It is a politics that seeks to institutionalize failure in public schooling through the implementation of increasingly impossible accountability regimes [eg., No Child Left Behind] in order to justify its colonization by private actors. It is a politics that seeks to carve out politicized markets from a "failing" public institution, markets that represent a direct conduit for private interests to tap public treasuries.[37]

The political characteristics of this articulation of power clearly emerged in the preceding inquiry. It is a politics that above all rejects ambiguity. The constellation of social forces, institutions, cultural practices and economic formations always already present in public schooling and individual student achievement is distilled down by this political bloc into that most transcendent of Enlightenment fetishes: numbers.[38] Indeed, it is a politics that demonstrates an instrumentality of reason borne of business plans and MBA curricula the world over, and it renders school learning down to the simple act of production in the form of test scores that are easily plotted on charts and presented in digital slideshows. That business leaders would formulate a model of schooling built on clearly defined goals, quantitative metrics, and punitive accountability measures should come as no surprise to anyone.[39] Yet, like all politics, it is a politics that is necessarily mediated through normative concepts at work in popular discourse and dominant ideological formations, concepts that are themselves battlegrounds in a 'war of position' among political actors vying for hegemonic dominance.[40] It is a politics that is not only contested but is a politics that creates the logical framework for its own critique and transformation. It opens itself up to its own colonization. Thus, we find in the logical framework of domination a logic of *potential* liberation.

As we have seen, the concept of accountability is being employed by this articulation of power to make specific epistemic claims that establish criteria for judging the validity of those claims, an epistemic test that current accountability policies fail. However, the critical moment consists of more than simply identifying epistemic failure or "contradiction." Critique must also work through the logic of the concepts employed by this articulation of power in order to both bring contradictions to the surface of public discourse and to open up a political space for the transformative negation of those concepts, or to establish the foundation for a positive reconstruction of those concepts through an articulation of the necessary pre-conditions for achieving their own normative ideals.

The results of this study suggest that the necessary pre-conditions for achieving the normative ideals built into the conceptual logic of accountability would look dramatically different from the accountability policies associated with standards-based reforms. The foundation for achieving the goals of preparing students for the intellectual rigors of a 21st century economy begins with a re-commitment to public schooling as a public institution in which all share responsibility and in which accountability pressures are centered around the ballot box. Upon this foundation, the first step toward achieving the normative ideals implicit within the conceptual framework of accountability is a coordinated effort to formulate academic standards that embrace the ambiguity of the complex knowledge we seek to impart to our students. The second step requires a coordinated effort to re-orient assessment and accountability away from punitive measures that creative perverse incentive structures that punish the weakest in society. Instead, what is needed is a movement toward formative assessments that both do justice to the complex knowledge we seek to impart to our students and are geared toward increasing student achievement and learning. And, finally, it demands an institutional structure of shared accountability among all educational actors. It demands that society re-affirm its commitment to schooling as a public institution.

NOTES

1 "Economic Security and a 21st Century Education: Secretary Arne Duncan's Remarks at the U.S. Chamber of Commerce's Education and Workforce Summit," *Speeches and Testimony*, http://www.ed.gov/news/speeches/2009/11/11092009.html (Accessed: November 10, 2009).

2 Of course, this economic perspective is not unique to the Obama administration. Popular debates around the relationship between global economic competition and public education can be traced back to the publication of *A Nation at Risk* during the Reagan administration.

3 Scott Ellison, "In the Shadow of Hegel: Toward a Methodology Appropriate to the Sociological Consciousness of Philosophic Inquiry," *Education and Culture, 26*(1), (2010): 44–66.

4 John Dewey, *How We Think* (New York: D.C. Heath & Co., 1910), 79.

5 Michel Foucault, *The Politics of Truth* (2nd ed.). (Los Angeles, CA: Semiotext(e), 2007), 114.

6 Ellison, 61.

7 Kevin Kosar, *Failing Grades: The Federal Politics of Education Standards* (Boulder, CO: Lynne Rienner Publishers, 2005), 47–48.

8 David T. Kearns & James Harvey, *A Legacy of Learning: Your Stakes in Standards and New Kinds of Public Schools* (Washington, DC: Brookings Institution Press, 2000), 139.

9 Kosar, 68.

10 Herbert J. Walberg, "Standards, Testing, and Accountability," in John E. Chubb (Ed.), *Within Our Reach: How America Can Educate Every Child* (Lanham, MD: Rowman & Littlefield Publishers, 2005), 55.

11 Frederick M. Hess, *Common Sense School Reform* (New York: Palgrave MacMillan, 2004), 51.

12 Finbarr C. Sloan & Anthony E. Kelly, "Issues in High-Stakes Testing Programs," *Theory Into Practice, 42*(1), (2003): 12–17.

13 Sloan & Kelly, 15.

14 Sloan & Kelly, 13.

15 Robert L. Linn, "Validity of Inferences from Test-Based Educational Accountability Systems," *Journal of Personnel Evaluation in Education, 19*(1–2), (2006): 14.

16 "Race to the Top Education Reforms Should Be Evaluated," National Academy of Sciences: The Office of News and Public Information of the National Academies, http://www8.nationalacademies.org/onpinews/newsitem.aspx?RecordID=12780# (Accessed: October 7, 2009).

17 Andrew Dean Ho, "The Problem With 'Proficiency': Limitations of Statistics and Policy Under No Child Left Behind," *Educational Researcher, 37*(6). (2008): 351–360.

18 Andre A. Rupp, Tracy Ferne, & Hyeran Choi, "How Assessing Reading Comprehension with Multiple Choice Questions Shapes the Construct: A Cognitive Processing Perspective," *Language Testing, 23*(4), (2006): 470.

19 Andre A. Rupp & Nonie K. Lesaux, "Meeting Expectations? An Empirical Investigation of a Standards-Based Assessment of Reading Comprehension," *Educational Evaluation and Policy Analysis, 28*(4), (2006): 319.

20 Rupp & Lesaux, 331.

21 Marsha Riddle Bully & Sheila W. Valencia, "Below the Bar: Profiles of Students Who Fail State Reading Assessments," *Educational Evaluation and Policy Analysis, 24*(3), (2002): 232.

22 Bully & Valencia, 233.

23 Sharon L. Nichols, Gene V. Glass, & David C. Berliner, "High-Stakes Testing and Student Achievement: Does Accountability Pressure Increase Student Learning?" *Educational Policy Analysis Archives, 14*(1), (2006): 51.

24 Gregory J. Merchant, Sharon E. Paulson, & Adam Shunk, "Relationships between High-Stakes Testing Policies and Student Achievement after Controlling for Demographic Factors in Aggregated Data," *Educational Policy Analysis Archives, 14*(30), (2006): 1–31.

25 Jaekyung Lee & Kenneth K. Wong, "The Impact of Accountability on Racial and Socioeconomic Equity: Considering Both School Resources and Achievement Outcomes," *American Educational Research Journal, 41*(4), (2004): 797–832.

26 Merchant, Paulson, & Shunk, 23. See also: Sharon Nichols & David Berliner, "Testing the Joy out of Learning," *Educational Leadership, 65*(6), (2008): 14–18.; Sharon Nichols, "High-Stakes Testing: Does It Increase Achievement?" *Journal of Applied School Psychology, 23*(2), (2007): 47–64.; Jaekyung Lee, "Trick or treat: new ecology of education accountability system in the USA," *Journal of Education Policy, 25*(1), (2010): 73–93.

27 Sharon L. Nichols & David C. Berliner, *Collateral Damage: How High-Stakes Testing Corrupts America's Schools* (Cambridge, MA: Harvard Education Press, 2007); Wayne Au, "High-Stakes Testing and Curricular Control: A Qualitative Metasynthesis," *Educational Researcher, 36*(5), (June 1, 2007): 258–267.

28 Juilian Vasquez Heilig & Linda Darling-Hammond, "Accountability Texas-Style: The Progress and Learning of Urban Minority Students in a High-Stakes Testing Context," *Educational Evaluation and Policy Analysis, 30*(2), (2008): 75–110; Linda McSpadden McNeil, Eileen Coppola, Judy Radigan, & Julian Vasquez Heilig, "Avoidable Losses: High-stakes accountability and the drop-out crisis," *Education Policy Archives,16* (2008): http://epaa.asu.edu/ojs/article/view/28/154; Brian A. Jacob, "Accountability, incentives and behavior: the impact of high-stakes testing in the Chicago Public Schools," *Journal of Public Economics 89* (2005): 763; Randall Reback, "Teaching to the rating: School accountability and the distribution of student achievement," *Journal of Public Economics 92*(5–6), (2008): 1412; Michelle Fine, Reva Jaffe-Walter, Pedro Pedraza, Valerie Futch, & Brett Stoudt, "Swimming: On Oxygen, Resistance, and Possibility for Immigrant Youth Under Seige," *Anthropology & Education Quarterly, 38*(1), (2007): 76–96; Kris Sloan, "High-Stakes Accountability, Minority Youth, and Ethnography: Assessing the Multiple Effects," *Anthropology & Education Quarterly 38*(1) (2007): 24–41; Linda Valli & Marilyn Chambliss, "Creating Classroom Cultures: One Teacher, Two Lessons, and a High-Stakes Test," *Anthropology & Education Quarterly 38*(1), (2007): 57–75; Linda Valli & Daria Buese, "The changing roles of teachers in an era of high-stakes accountability," *American Educational Research Journal, 44*(3), (September 2007): 519–558.

29 Barbara J. Thayer-Bacon, *Transforming Critical Thinking: Thinking Constructively* (New York: Teachers College Press, 2000).

30 P. Sahlberg, "Education Reform for Raising Economic Competitiveness," *Journal of Educational Change, 7*(4), (December 1, 2006): 259–287; Tiina Itkonen & Markku Jahnukainen, "An Analysis of Accountability Policies in Finland and the United States," *International Journal of Disability, Development and Education, 54*(1), (2007): 5; Andreas Schleicher, *The Economics of Knowledge: Why Education is Key to Europe's Success* (Brussels, Belgium: The Lisbon Council, 2006), http://www.lisboncouncil.net/index.php?option=com_publications&task=view&id=1&Itemid=35; *OECD Briefing Note For Finland* (OECD, September 9, 2008), http://www.oecd.org/dataoecd/31/46/41277828.pdf.

31 OECD, *Program for International Student Assessment* (2009) www.oecd.org/dataoecd/51/27/37474503.pdf

32 Pasi Sahlberg, "Education Policies for Raising Student Learning: The Finnish Approach," *Journal of Education Policy, 22*(22), (2007): 147–171.

33 Pasi Sahlberg, "Rethinking accountability in a knowledge society," *Journal of Educational Change, 11*(1), (2008): 55.

34 Linda Darling-Hammond, "Standards, Accountability, and School Reform," *Teachers College Record, 106*(6), (2004): 1047–1085.

35 Elizabeth DeBray-Pelot & Patrick McGuinn, "The New Politics of Education: Analyzing the Federal Education Policy Landscape in the Post-NCLB Era," *Educational Policy, 23*(1), (January 1, 2009): 15–42; Janelle Scott, "The Politics of Venture Philanthropy in Charter School Policy and Advocacy," *Educational Policy, 23*(1), (January 1, 2009): 106–136; Christopher Lubienski, Peter Weitzel, & Sarah Theule Lubienski, "Is There a "Consensus" on School Choice and

Achievement?: Advocacy Research and the Emerging Political Economy of Knowledge Production," *Educational Policy, 23*(1), (January 1, 2009): 161–193.

36 Rob Roberts, "Entertainment Properties Trust invests $48M in charter schools," *Kansas City Business Journal*, January 22, 2010, http://www.bizjournals.com/kansascity/stories/2010/01/18/daily54.html

37 John Clarke, "Dissolving the Public Realm? The Logics and Limits of Neo-liberalism," *Journal of Social Policy, 33*(1), (2004): 30; Michael Apple, "Creating Difference: Neo-Liberalism, Neo-Conservatism and the Politics of Educational Reform," *Educational Policy, 18*(1), (2004): 12–44; Michael Apple, "Between Neoliberalism and Neoconservatism: Education and Conservatism in a Global Context," in Nicholas C. Burbules & Carlos Alberto Torres, (Eds.), *Globalization and Education: Critical Perspectives* (New York: Routledge, 2000).

38 Max Horkheimer & Theodor W. Adorno, *Dialectic of Enlightenment* (1st ed.). (Palo Alto, CA: Stanford University Press, 2007).

39 Indeed, the factory model still with us today is itself a product of the industrial era from which it emerged.

40 Antonio Gramsci, *Prison Notebooks, Volume 2* (New York: Columbia University Press, 1996), 200–201, 156–158, 52–53.

REFERENCES

Apple, M. (2004). Creating difference: Neo-liberalism, neo-conservatism and the politics of educational reform. *Educational Policy, 18*(1), 12–44.

Apple, M. (2000). Between neoliberalism and neoconservatism: Education and conservatism in a global context. In N. C. Burbules & C. A. Torres (Eds.), *Globalization and education: Critical perspectives*. New York: Routledge.

Au, W. (2007). High-stakes testing and curricular control: A qualitative meta-synthesis. *Educational Researcher, 36*(5), 258–267.

Bully, M. R., & Valencia, S. W. (2002). Below the bar: Profiles of students who fail state reading assessments. *Educational Evaluation and Policy Analysis, 24*(3), 219–239.

Clarke, J. (2004). Dissolving the public realm? The logics and limits of neo-liberalism. *Journal of Social Policy, 33*(1), 27–48.

Darling-Hammond, L. (2004). Standards, accountability, and school reform. *Teachers College Record, 106*(6), 1047–1085.

DeBray-Pelot, E., & McGuinn, P. (2009). The new politics of education: Analyzing the federal education policy landscape in the post-NCLB era. *Educational Policy, 23*(1), 15–42.

Dewey, J. (1910). *How we think*. New York: D. C. Heath & Co.

Ellison, S. (2010). In the shadow of Hegel: Toward a methodology appropriate to the sociological consciousness of philosophic inquiry. *Education and Culture, 26*(1), 44–66.

Fine, M., Jaffe-Walter, R., Pedraza, P., Futch, V., & Stoudt, B. (2007). Swimming: On oxygen, resistance, and possibility for immigrant youth under seige. *Anthropology & Education Quarterly, 38*(1), 76–96.

Foucault, M. (2007). *The politics of truth* (2nd ed.). Los Angeles: Semiotext(e).

Gramsci, A. (1996). *Prison notebooks, Volume 2*. New York: Columbia University Press.

Heilig, J. V., & Darling-Hammond, L. (2008). Accountability Texas-style: The progress and learning of urban minority students in a high-stakes testing context. *Educational Evaluation and Policy Analysis, 30*(2), 75–110.

Hess, F. M. (2004). *Common sense school reform.* New York: Palgrave MacMillan.

Ho, A. D. (2008). The problem with "proficiency": Limitations of statistics and policy under No Child Left Behind. *Educational Researcher, 37*(6), 351–360.

Horkheimer, M. & Adorno, T. W. (2007). *Dialectic of enlightenment* (1st ed.). Palo Alto, CA: Stanford University Press.

Itkonen, T., & Jahnukainen, M. (2007). An analysis of accountability Policies in Finland and the United States. *International Journal of Disability, Development, and Education, 54*(1), 5–23.

Jacob, B. A. (2005). Accountability, incentives, and behavior: The impact of high-stakes testing in the Chicago Public Schools. *Journal of Public Economics, 89,* 761–796.

Kearns, D. T., & Harvey, J. (2000). *A legacy of learning: Your stakes in standards and new kinds of public schools.* Washington, DC: Brookings Institution Press.

Kosar, K. (2005). *Failing grades: The federal politics of education standards.* Boulder, CO: Lynne Rienner Publishers.

Lee, J. (2010). Trick or treat: new ecology of education accountability system in the USA. *Journal of Education Policy, 25*(1), 73–93.

Lee, J., & Wong, K. W. (2004). The impact of accountability on racial and socioeconomic equity: Considering both school resources and achievement outcomes. *American Educational Research Journal, 41*(4), 797–832.

Linn, R. L. (2006). Validity of inferences from test-based educational accountability systems. *Journal of Personnel Evaluation in Education, 19*(1–2), 5–15.

Lubienski, C., Weitzel, P., & Lubienski, S. T. (2009). Is there a "consensus" on school choice and achievement?: Advocacy research and the emerging political economy of knowledge production. *Educational Policy, 23*(1), 161–193.

McNeil, L. M., Coppola, E., Radigan, J., & Heilig, J, V. (2008). Avoidable losses: High-stakes accountability and the dropout crisis. *Education Policy Archives, 16,* http://epaa.asu.edu/ojs/article/view/28/154

Merchant, G. J., Paulson, S. E., & Shunk, A. (2006). Relationships between high-stakes testing policies and student achievement after controlling for demographic factors in aggregated data. *Educational Policy Analysis Archives, 14*(30), 1–31.

Nichols, S. (2007). High-stakes testing: Does it increase achievement? *Journal of Applied School Psychology, 23*(2), 47–64.

Nichols, S. & Berliner, D. (2008). Testing the joy out of learning. *Educational Leadership, 65*(6), 14–18.

Nichols, S, & Berliner, D. C. (2007). *Collateral damage: How high-stakes testing corrupts America's schools.* Cambridge, MA: Harvard Education Press.

Nichols, S., Glass, G. V., & Berliner, D. C. (2006). High-stakes testing and student achievement: Does accountability pressure increase student learning? *Educational Policy Analysis Archives, 14*(1), 1–175.

Reback, R. (2008). Teaching to the rating: School accountability and the distribution of student achievement. *Journal of Public Economics, 96*(5–6), 1394–1415.

Rupp, A. A., Ferne, T., & Choi, H. (2006). How assessing reading comprehension with multiple choice questions shapes the construct: A cognitive processing perspective. *Language Testing, 23*(4), 441–474.
Rupp, A. A., & Lesaux, N. K. (2006). Meeting expectations? An empirical investigation of a standards-Based assessment of reading comprehension. *Educational Evaluation and Policy Analysis, 28*(4), 315–333.
Sahlberg, P. (2008). Rethinking accountability in a knowledge society. *Journal of Educational Change, 11*(1), 45–61.
Sahlberg, P. (2007). Education policies for raising student learning: The Finnish approach. *Journal of Education Policy, 22*(22), 147–171.
Sahlberg, P. (2006). Education reform for raising economic competitiveness. *Journal of Educational Change, 7*(4), 259–287.
Schleicher, A. (2006). *The economics of knowledge: Why education is key to Europe's success*. Brussels, Belgium: The Lisbon Council.
Scott, J. (2009). The politics of venture philanthropy in charter school policy and advocacy. *Educational Policy, 23*(1), 106–136.
Sloan, K. (2007). High-stakes accountability, minority youth, and ethnography: Assessing the multiple effects. *Anthropology & Education Quarterly, 38*(1), 24–41.
Sloan, F. C., & Kelly, A. E. (2003). Issues in high-stakes testing programs. *Theory Into Practice, 42*(1), 12–17.
Thayer-Bacon, B. J. (2000). *Transforming critical thinking: Thinking constructively*. New York: Teachers College Press.
Valli, L., & Buese, D. (2007). The changing roles of teachers in an era of high-stakes accountability. *American Educational Research Journal, 44*(3), 519–558.
Valli, L., & Chambliss, M. (2007). Creating classroom cultures: One teacher, two lessons, and a high-stakes test. *Anthropology & Education Quarterly, 38*(1), 57–75.
Walberg, H. J. (2005). Standards, testing, and accountability. In J. E. Chubb, *Within our reach: How America can educate every child*. Lanham, MD: Rowman & Littlefield.

Reading 1.2

GUIDELINES FOR IMPROVING ACCOUNTABILITY SYSTEMS

Laura S. Hamilton and Brian M. Stecher

TREAT A TEST-BASED ACCOUNTABILITY SYSTEM AS DYNAMIC AND FLEXIBLE

Change test items regularly so that individual items do not become too familiar to teachers and test-takers.

Consider major system upgrades to take advantage of new designs or enhancements when past practices become obsolete.

Laura S. Hamilton and Brian M. Stecher, Selection from "Improving Test-Based Accountability," *Making Sense of Test-Based Accountability in Education*, ed. Laura S. Hamilton, Brian M. Stecher, and Stephen P. Klein, p. 124.

MONITOR AND IMPROVE TECHNICAL QUALITY

Conduct ongoing evaluation of technical quality to ensure that the systems continue to meet the needs of students, educators, policymakers, and the public.

Incorporate multiple measures and multiple formats to reduce the risk of making incorrect decisions.

Incorporate findings from cognitive psychology to ensure that tests reflect (1) the way students learn and demonstrate competence and (2) the system's instructional goals.

Collect and maintain rich student-level data that allow tracking individual students over time.

Adopt stable statistical indicators that reflect real differences in student achievement and minimize measurement and sampling errors.

DESIGN SYSTEMS THAT HAVE POSITIVE EFFECTS ON PRACTICE

Align teacher professional development and curriculum with standards and assessments so that teachers understand and support the system's goals and respond with effective classroom practices.

Conduct ongoing research and evaluation of effects on practice to guide school reform efforts.

Clarify the desired balance between accountability purposes and instructional purposes as a basis for making decisions among competing demands.

Create incentives that promote desired goals, thus taking advantage of the strong response that high-stakes testing elicits from schools and teachers.

Consider each policy's potential effects on equity to improve the performance and opportunities of students who have traditionally experienced low levels of academic success.

CONSIDER THE POLITICAL CONTEXT

Facilitate integration of political and professional perspectives to remove obstructions to progress and build better tests and accountability systems.

Incorporate parents' perspectives to make sure that accountability systems provide the information needed for more-effective parental involvement and decisionmaking.

Explore costs and benefits of alternative approaches in order to make well-informed decisions.

HOW DO ASSESSMENT AND ACCOUNTABILITY APPLY TO MY PROFESSION?

Assessment is one of the most important things that educators can do to help students learn. Assessment can actually drive student learning, but students are often confused about what is asked of them regarding assessment. If students are engaged in the entire learning process, learning will be positively impacted. The various aspects of the learning process should take into account pertinent dialogue concerning the development and use of assessments. Through this engagement, students will be allowed to construct their own meanings, thereby providing a more effective understanding of the subject matter at hand. A more effective understanding of a concept or process induces deeper learning and personal ownership of learning. When student learning and assessment are viewed as inseparable components of a dynamic process, the ongoing development and continual improvement of instructional practices will naturally occur. A problem, however, is that teachers often make instructional decisions apart from assessment data. This can diminish the intended uses of assessment as a diagnostic or formative tool to improve teaching and learning. So, instead, it's imperative that as a classroom teacher or aspiring leader, you are explicit and transparent about the purpose of the assessment, the data that can be obtained as a result of the assessment, and how the data will be used to inform decisions about teaching and student learning.

CHAPTER 2

Assessment Versus Accountability Versus Measurement Versus Evaluation Versus Research

What do they all mean and why should I care?

AS YOU READ

As you read the chapter, think about the following:

- How do **you** define or explain assessment versus accountability versus measurement versus evaluation versus research?
- What is **your** experience with each of these concepts?

WHAT ARE THE GENERAL PRINCIPLES OF ASSESSMENT VERSUS ACCOUNTABILITY VERSUS MEASUREMENT VERSUS EVALUATION VERSUS RESEARCH?

People often use the terms measurement and assessment interchangeably. Although there are some similarities, there are significant differences. This book will help you to become fluent in those differences and similarities. There is no one accepted definition of either of the two concepts. As you read in Chapter 1, assessment serves many purposes and your conception of assessment is based on your experiences, attitudes, understandings, beliefs, and so on. In other words, the term assessment is contextual. In this text, when I refer to assessment, I am generally referring to how we collect data or the evidence we collect from our students or teachers. The purpose of assessing is the number one priority when determining how to assess. For example, if you want to determine a student's knowledge of a concept, you may choose to "test" the student. You test a student using extant or created instruments. There is a plethora of types of instruments that you may select or develop to test a student's knowledge of the concept. For example, you may choose to use an end-of-chapter test provided to you by the textbook company. Instead, you may think it is best to write your own questions that will be used to test the student's knowledge of the concept. We will take a critical look at each of these options.

In the first instance, you choose to use an end-of-chapter exam that is provided by the textbook that the district has selected for your school (grade level, subject-area, etc.). After all, you did have the students read most of the chapter, so the questions should align to their knowledge. You then administer that exam to test the student's knowledge of the concept. You then score the exam to determine the student's grade on that test that equals his knowledge—right? This is a common practice in classrooms today. You make the assumption that you are using a "good test" of a student's knowledge about the concept because a commercial company developed the questions. Well, there is a lot more to this that we will discuss in a later chapter.

In the second instance, you decide that you want to write your own questions that will comprise your created test. So, you diligently work on writing questions that you believe will generate responses equal to the student's knowledge of the concept. You think that writing your own questions is better than using the extant end-of-chapter exam because the questions will better align to your instruction. So, you then administer, score, and assign a grade on that test that equals his knowledge—right? You assume you are using a "good test" of students' knowledge about the concept because you created the questions. As is the case above, there is a lot more to this that we will discuss in a later chapter.

Based on these two examples, which approach is better—using the end-of-chapter exam or developing your own exam? The short answer is that it depends. In a later chapter, I will explain how to best make that decision.

Let's get back to the terms assessment and measurement. Earlier, I indicated that **assessment** is the means by which you collect data. Using these examples, the actual instruments themselves are the assessments (end-of-chapter exam and/or the created test). We use the term assessment in a general sense as synonymous with the instrument or test itself. In the psychometric world, many experts refer to the assessment instruments as "measures" or "measures of assessment."

What if we look at the term "assess" instead of "assessment"? For the purposes of this text, when we refer to the term assess, we are referring to what it is that we want to measure. In the example above, we want to assess (determine, measure) a student's knowledge of a concept. Thus, we assess or measure that knowledge using an assessment instrument.

Accountability is not specific to the profession of education. It is more in the forefront now than it was decades ago; however, the concept is not new as you read in Chapter 1. In its simplest form, accountability is taking responsibility or liability for actions, decisions, and/or policies. Each action, each decision, and each policy have implications. Accountability helps internal and external stakeholders monitor and disclose the impact of those implications. Accountability is often the umbrella under which all of the following concepts fall.

Measurement is the term used when we talk about measuring something—in this case, we are trying to measure a child's knowledge. To measure the students' knowledge, we must create a scale for which to measure that knowledge. This is what I referred to earlier about how to determine which is better—creating your own test questions or using the end-of-chapter exam. There are ways that psychometricians determine "how good" a test is or "how-well" items align to what it is you are trying to measure. There are multiple strategies that are employed to determine the quality of the test questions and overall test. This book will cover many of those strategies, so you can become an assessment and measurement expert—whether you are a classroom teacher, school-wide leader, or district coordinator.

Evaluation typically includes the process by which one makes value judgments. In education, programs and curricula are evaluated. Evaluation refers to the effectiveness of a curriculum or program. The meaning of the term differs from assessment in that it is product focused as opposed to diagnostic in nature.

Research refers to the study, investigation, or examination of one or more variables to discover or interpret facts. It is empirical in nature. This term differs from assessment, measurement, and evaluation in that the process should be nonbiased. Specifically, a primary purpose of research is to develop new knowledge—not confirm what you already believe to be true. You may hypothesize about relationships, variables, causes, effects, etc., based on past empirical evidence. However, opinions are not imposed in research.

Each of these terms have similar yet different meanings and it is important to be able to distinguish between them in your own work. The purpose of each is driven by the outcomes you are trying to achieve. When you assess a child's knowledge, it is important that you use appropriate tools to assess his knowledge. Psychometrics are key in measurement—they help you determine the quality of the assessment tools that you use to make conclusions about student knowledge. When making judgments about a curriculum and/or program that has been implemented, it is important to follow the guidelines set forth for effectively evaluating the curriculum and/or program. Finally, when creating new knowledge about variables, you use research processes to collect the necessary data that elicit results from which you can make conclusions. The outcomes you are seeking will drive whether you employ strategies from assessment, measurement, evaluation, or research.

WHAT DOES THE LITERATURE SAY ABOUT ACCOUNTABILITY, ASSESSMENT, MEASUREMENT, EVALUATION, AND RESEARCH?

Reading 2.1

INTRODUCTION TO MEASUREMENT

Malcolm L. van Blerkom

INTRODUCTION

In this [reading] we will begin to look at the various ways in which we use measurement in the classroom. We will differentiate between the terms assessment, measurement, and evaluation as well as differentiate between formal and informal assessment. We will look at the various roles of assessment including preliminary, diagnostic, formative, and summative assessment. We will also differentiate between maximum and typical performance measures. [...]

THE ROLE OF MEASUREMENT

Measurement plays many different roles in our lives. During the past week I have had to rely on measurement in many different ways. The other day, while at the grocery store, I had to locate a can of diced tomatoes that was 14.5 ounces for a recipe for Spanish rice. Each morning I rely on my alarm clock to wake me at 6:45 a.m. (I get to sleep-in late this year!) I have recently anxiously watched the price for a gallon of gasoline rise and fall. When a student asked me for a recommendation, I reviewed his performance in my class by checking my grade book, paying special attention to his test scores. These examples represent only a few of the many ways that I use measurement on a daily basis.

You will also use measurement a great deal in your classrooms. You may give students diagnostic tests to assess how well they are reading, spelling, writing, counting, and so on. You will give tests to assess how well your students have learned and mastered the material that you have been teaching. You will use informal assessment on a daily basis to judge how well your lessons are progressing: Do your students appear to be comprehending your instruction? You will collect homework to assess if your students are actually able to apply the new skills that they have been learning. As you will see, measurement will play many different roles in our classrooms.

In [another section] we discuss statistics. That discussion is designed primarily to give you sufficient background to be able to examine and comprehend the many measurement concepts that we will be discussing [...]. Those statistical concepts that will be most often used in the discussions of measurement include the mean, the standard deviation, the variance, correlation, and z-scores. You will see how we use these and other statistical concepts to build good measurement practices.

MEASUREMENT, ASSESSMENT, AND EVALUATION

There are many terms that describe the processes that we use to judge student performance. The three terms most commonly used are measurement, assessment, and evaluation. Although I will provide you with specific definitions for each term, I must warn you that there is no international law or agreement on how these terms should be used. I have seen many instances where these three terms are freely used interchangeably as if they all meant exactly the same thing. However, I will give you specific definitions for each according to the ways that each term is used most frequently in education.

Assessment

The most general term is assessment. *Assessment* is a very general term that describes the many techniques that we use to measure and judge student behavior and performance. Although in some professional circles the term assessment sometimes means something more specific, in relation to the classroom, it is typically a very general, very generic term. When in doubt about how to label a technique, the safest label to use would be assessment.

Measurement

Over the years, I have seen many definitions of measurement. However, here is the one that I find the most useful. *Measurement* is the process of assigning meaningful numbers (or labels) to persons or objects based on the degree to which they possess some characteristic.

Let's look at some examples and start with how we could take measurements of an object. What are some characteristics of a classroom to which we could assign meaningful numbers or labels? The first thing that we generally think of is the classroom's dimensions, especially its length and width. A second important characteristic is the classroom's seating capacity. Some classrooms may be designed to hold 20 students, whereas others could accommodate 35 to 40 students. This becomes important when we are assigning classes to certain rooms. Still another characteristic that I find important is board space. I tend to use the board a great deal, especially in some classes, and want classrooms with plenty of board space (perhaps measured in square feet). What about the room number? For example, let's say that we are in room 231. Is the room number a form of measurement? It may be! The room number is frequently used as a substitute for a name and can be considered measurement at the nominal (naming) level [...]. If room numbers are assigned to rooms randomly and don't assist a stranger in finding the room, then I would argue that in that case the room number was not a form of measurement. In that case the room number was not very meaningful. However, if room numbers are assigned in some logical order (as they are in most buildings) and they assist a stranger in finding the room, then I would argue that they are measurement, if only at its least sophisticated level. In this case the room number has some meaning and tells you something about the classroom, where it is located.

For another example, let's use a student. What are some characteristics of a student that can be described with meaningful numbers or labels? Of course, many of the student's various physical characteristics such as height, weight, and age can all be described with meaningful numbers. Even describing other characteristics such as gender are frequently considered meaningful at the nominal level. Other characteristics to which we can ascribe meaningful numbers would include the student's score on the last social studies quiz, the student's score on an I.Q. test, or her score on the standardized math achievement test. There are many student characteristics that we can measure in a meaningful way.

Measurement is primarily a mechanical process. When we score a test, count up the number of points that the student earned, and record that number, we are using measurement.

Evaluation

The third term is evaluation. *Evaluation* involves the use of measurement to make decisions about or to determine the worth of a person or object. Evaluation is frequently the step that follows measurement. Let's say that Dana earned 87 on a 100-point science test. When we scored Dana's test and determined that she earned 87 out of 100 points we were using measurement. It was a mechanical process. If we then decide that Dana's test score of 87 translates to a letter grade of B, we are now using evaluation. Deciding on a grade was dependent on a judgment of worth. A letter grade is typically a statement of worth: A is excellent, B is good, C is average, and so on. So, scoring a test is measurement, but assigning a grade is evaluation.

Evaluation comes in many forms. One obvious form is the use of letter grades. However, we sometimes run courses on a pass/fail basis. If your mean test score is 75% or higher, you pass. If it is below 75% you fail. At other times teachers may prepare a narrative statement concerning how well the student is performing and use terms like "making satisfactory progress," or "still needs more work in this area." These are also forms of evaluation.

Definitions

Assessment is a very general term that describes the many techniques that we use to measure and judge student behavior and performance.
Measurement is the process of assigning meaningful numbers (or labels) to persons or objects based on the degree to which they possess some characteristic.
Evaluation involves the use of measurement to make decisions about or to determine the worth of a person or object.

Reading 2.2

FURTHER STEPS TOWARD THE DEVELOPMENT OF AN ACCOUNTABILITY-ORIENTED SCIENCE OF MEASUREMENT

Daniel Koretz

Large-scale educational assessment has gone through many changes in recent decades, but none has implications for the practice of measurement so great as the shift to using tests as a means of holding educators and educational systems accountable. Achievement testing has never been entirely free of this purpose, but the balance of goals has shifted greatly, and accountability has become the most important function of many large-scale testing programs in the United States (Koretz & Hamilton, 2006) and increasingly in other nations as well.

Research has brought to light many serious concerns about the functioning and effects of test-based accountability systems. Yet the science and practice of measurement have been slow to respond, continuing in key respects much as they had before the shift to accountability-oriented testing. The consequences of this inertia are serious, including biased measurement and distorted incentives for educators.

To clarify the issues raised by test-based accountability, one can think of the evolution of modern measurement as having three stages. This is an oversimplification; the dividing lines among the stages are fuzzy, and one can find in each stage early examples of the types of work that characterize the later ones. Nonetheless, the distinctions are useful in that they highlight both recent developments related to accountability-oriented testing and challenges that remain.

The entire history of modern measurement up to the 1980s constitutes the first stage, which could be called "traditional psychometrics." High-stakes testing was not uncommon during this era. (The term "high stakes" is often used to refer only to tests that have tangible consequences for students or teachers. However, the term is used here to refer to tests on which teachers or students feel pressured to raise scores, even if the mechanism of pressure is not concrete sanctions or rewards.) Indeed, high-stakes testing existed long before the science of measurement began. Nonetheless, traditional psychometrics was in two critical respects tacitly premised on low stakes. The first is that it gave relatively little attention to the consequences of testing. The second is a special case of the first: traditional psychometrics focused little on behavioral responses to testing, other than the behavior of the student while taking the test and of proctors administering it. Most important for present purposes, traditional psychometrics gave little attention to behavioral responses to testing that might threaten the validity of score-based inferences about student achievement.

In the traditional view, changes in behavior in response to testing were expected, and indeed one of the functions of testing has long been to foster them. For instance, a common rationale for the norm-referenced tests that dominated achievement testing in the United States until the end of the period of traditional psychometrics was the expectation that teachers would use the information the test provided about the relative strengths and weaknesses of students' performance to refine their instruction. But a key, if sometimes tacit, assumption was that if students' performance on a test improved because of such responses to testing, this change represented real improvements in achievement, not test-specific gains in scores. For this reason, these behavioral responses were of little concern to psychometrics and did not greatly complicate decisions about design, methods, or validation.

Of course, warnings that this assumption was unrealistic were voiced long ago. For example, Lindquist offered this caution in a standard reference in 1951:

> The widespread and continued use of a test [that is a proxy for an unattainable measure of criterion behaviors] will, in itself, tend to reduce the correlation between the test series and the criterion series for the population involved. Because of the nature and potency of the rewards and penalties associated in actual practice with high and low achievement test scores of students, the behavior measured by a widely used test tends in itself to become the real objective of instruction, to the neglect of the (different) behavior with which the ultimate objective is concerned.
> (Lindquist, 1951, pp. 152–153)

Nonetheless it is fair to say that most of the psychometric enterprise—what people in the field did when developing methods or operating testing programs—proceeded without much attention to these concerns.

The labeling of this period as "traditional" is unrelated to the theoretical and mathematical models used in the construction of tests and the analysis of scores. It is not related to the conventional distinction between "classical" and "modern" approaches to measurement, that is, between classical test theory and both generalizability theory and item response theory (IRT). Although generalizability theory could be extended to address issues raised by accountability-oriented testing (e.g., to evaluate the generalizability of performance from tests used for accountability to other tests), it is not typically used in this way. Similarly, while it may be feasible to employ IRT as a tool for evaluating the results of accountability-oriented testing (e.g., Koretz & McCaffrey, 2005), IRT models are usually applied in a manner that assumes no general behavioral responses to testing that would undermine validity. Specifically, IRT retains the traditional assumption that changes in item-level performance most often reflect true changes in proficiency. Thus, the shift to modern test theory has not taken the field beyond traditional psychometrics in the sense that the latter term is used here.

During the 1970s and 1980s, the field of measurement entered a "traditional psychometrics plus" period that continues to the present. This stage was marked by important developments in both theory and empirical research relevant to the growing use of large-scale assessments for monitoring aggregate performance and for accountability. However, the actual practice of educational measurement, from test design through validation, has changed far less.

During this stage, the most important theoretical development bearing on the evolving uses of achievement tests was the growing attention to the effects of testing. The notion that validity encompasses consequences of testing as well as the reasonableness of the inferences based on scores dates back at least to the middle of the last century (Kane, 2006; Shepard, 1997), but it became much more central to the theory of validity over the past several decades (e.g., Linn, 1997; Messick, 1989; Shepard, 1997). Argument continues about whether it is useful to group both the impact of testing and the appropriateness of score-based inferences under the single term "validity" (e.g., compare Shepard, 1997 and Popham, 1997). For present purposes, however, these arguments are not critical, and the term validity is used below only to refer to the reasonableness of score-based inferences.

At the risk of oversimplification, one could say that most of the discussion of consequences in writings about validity has focused on the effects of testing as outcomes in their own right. For example, Shepard (1997, p. 8) notes concerns about medical students taking fewer courses in the humanities and social sciences because of the content of the MCAT exam and comments that "these consequences could hardly be brought under the tent of validity for judging the test battery as a measure of science knowledge."

However, a few writers began to focus on the risk that some behavioral responses to testing could bias inferences about achievement by creating larger gains in scores than real gains in achievement warrant, influencing the validity of the test. This bias is typically called "score inflation" or "score corruption." For example, Linn, Baker, and Dunbar (1991) wrote that

> considering validity in terms of consequences forces our attention on aspects of the assessment process that may not be intended or anticipated by the designers of the instruments. We know from experience that results from standardized tests can be corrupted [i.e., inflated], and we have clear examples of some of the factors that lead to that corruption. (p. 17)

The past several decades have also witnessed a growth in empirical research exploring the effects of accountability-oriented testing programs. A limited amount of work has investigated the validity of gains obtained under high-stakes conditions (e.g., Jacob, 2005, 2007; Koretz, Linn, Dunbar, & Shepard, 1991; Koretz & Barron, 1998). A somewhat larger body of research has used surveys and case study methods to explore the responses of teachers and principals to accountability-oriented testing programs (e.g., Stecher, 2002). Research has also explored numerous other components of test-based accountability systems, such as problems in the establishment and the use of performance standards (e.g., Burstein et al., 1995/1996; Linn, 2003; Shepard, 1994) and inconsistencies among alternative aggregate measures of performance (e.g., Linn, 2000).

This new theoretical and empirical work [...] has profound implications both for the practice of measurement and for the design of test-based accountability systems. For example, research has shown that the responses of teachers and principals to these systems include not only some desired effects, but also a variety of undesirable behaviors (e.g., excessive narrowing of instruction). These findings call into question the simple model of incentives that motivates this form of assessment. And more important for the science of measurement is the finding that scores can become seriously inflated, introducing a bias that is sometimes very large and that undermines the key inferences based on scores. That is, the traditional working assumption that behavioral responses to testing do not seriously threaten validity, only somewhat problematic in Lindquist's day, appears to have become untenable.

Yet, the main body of the measurement enterprise has been largely impervious to the implications of these theoretical and empirical developments and has continued largely as before—hence the label traditional psychometrics plus, which is intended to imply a veneer of accountability-related work alongside a much larger body of traditional psychometric work. This is true of the whole span of measurement activities, starting with test design, continuing through many of the nuts and bolts of the psychometric enterprise, such as linking methods, and ending with validation. If this seems to be an overstatement, search recent technical reports of operational assessments or the texts used to train students in measurement to see how much discussion is devoted to the issues raised here.

The theoretical and empirical work on accountability-oriented testing conducted over the past two decades suggests the need to move to a third stage, which might be called "accountability-oriented measurement." It is time to take more seriously the warning Lindquist offered more than half a century ago and to begin to adapt the core work of the measurement enterprise to the widespread use of tests for accountability. Of course, other uses of testing will continue, and even for accountability testing, all of the concerns of traditional psychometrics will remain critically important. Therefore, this third stage requires additions to current work, not its replacement. Which aspects of the practice of measurement will need modification remains arguable; the argument has barely been joined, and a great deal of research remains to be done. But, we already have enough evidence to suggest that changes should span the entire process, starting with test design and continuing through validation.

ISSUES ARISING FROM THE USE OF TESTS FOR ACCOUNTABILITY

There is not sufficient space here to review all of the research exploring the implications of accountability-oriented testing. However, it is necessary to sketch a number of issues in order to explore their possible implications for measurement practice.

The issues raised to date can be categorized as behavioral and non-behavioral, with behavioral issues being those that arise when behavioral responses to testing affect test validity. The distinction is not always clean because some issues—for example, the reliance on standards-based reporting—have both behavioral and non-behavioral aspects, but it is conceptually important. Consistent with its history, the field of measurement has been more responsive to the non-behavioral issues, but the behavioral issues may have even more profound implications for measurement practice.

Non-Behavioral Issues

Error

Accountability-oriented testing has altered the problem of error. The discussion of error in traditional psychometrics focuses primarily on measurement error that is, inconsistencies across repeated measurements of a single individual, such as variations arising from the sampling of test items (e.g., individual differences from one form of the SAT to another) or from fluctuations in a given student's performance over time (e.g., from having more sleep before retaking an exam). In classical test theory, measurement error takes the form of simple random inconsistencies, while generalizability theory considers both random and systematic sources of measurement error. Sampling error—that is, inconsistencies arising from sampling of units of analysis, whether these are individual students or groups—received only limited attention because the science of measurement focused primarily on the precision of estimates of attributes of individuals. For the most part, sampling error remained the province of statisticians and analysts employing test scores for other purposes.

But because of accountability-oriented testing, sampling error has become more important, and the field has responded quickly to this change. Many of the most important results of current large-scale testing programs are aggregate scores, and the inferences based on these aggregate scores are often about classrooms, schools, or systems, not about the particular cohorts of students who produced them. Thus, one element of traditional psychometrics plus has been a growing attention to sampling error. For example, Kane and Brennan (1977) extended the use of generalizability analysis to evaluating the reliability of class means. They argued that this extension was important because of large-scale program evaluations that used classes as the unit of analysis and "studies of school effectiveness and accountability" (p. 267).

Development of generalizability analysis to address aggregate statistics that include sampling error continues. For example, Cronbach, Linn, Brennan, & Haertel (1997) explored the application of generalizability analysis to performance assessments as indicators of school effectiveness. Brennan, Yin, & Kane (2003) investigated its application to both norm-referenced and criterion-referenced inferences about group means and explored the application of multivariate generalizability analysis to data from adjacent grades.

In addition to developments of generalizability theory, recent work on sampling error has included research on error estimates for the percent of students scoring above a cut score, on the instability of aggregate results over time, and on sources of aggregate instability apart from simple sampling error. No Child Left Behind requires that performance be reported for aggregates smaller than entire schools (e.g., for each school's population of students with disabilities, if there are enough of them to produce reliable estimates) and this has led to considerable debate about the instability of the results of this disaggregated reporting (e.g., Kane & Staiger, 2002).

Standards-Based Reporting

The shift to accountability-oriented testing need not have been accompanied by a change to standards-based reporting, but in practice it was, perhaps because both accountability testing and standards-based reporting reflect a focus on expectations for performance rather than on descriptions of the current distribution of performance. Research investigating this change in reporting has raised concerns about the adequacy of the standards themselves (e.g., their robustness, their reasonableness given normative data, the appropriateness of and justification of the verbal labels they are given), the adequacy of standards-based statistics as a summary of performance, and the behavioral incentives created by reporting results only when they reflect a move from one performance category to another (e.g., from basic to proficient).

Despite often sharing similar or identical verbal labels, performance standards are highly inconsistent from state to state (Braun & Qian, 2007; Linn, 2003). And, they are often set at levels that appear unreasonable given normative data (Linn, 2000). Perhaps more troubling than these problems, which in theory can be remedied, is evidence that the results of standard setting are not robust and that variations among standard-setting results can be large. This has been clear at least since Jaeger's review of the literature through the 1980s. For example, Jaeger examined the percentages labeled as failing by different methods in 32 published comparisons and found that the median ratio of the largest to the smallest was 1.5 and the mean 5.3 (Jaeger, 1989). Shepard (1994) showed that the results of item-based and holistic methods are inconsistent, and in the case of the Angoff methods used with NAEP, are not robust across item formats or levels of difficulty. Linn (2003) noted that the composition of panels of judges also introduces inconsistencies, although he warned that the data on this point remain limited. In a recent article, Linn summarized empirical evidence about performance standards and argued against using this form of reporting. He maintained that "the variability [in state standards] is so great that characterizing achievement is meaningless" and argued that "There are a number of reasons to question the wisdom of setting a performance standard for a test if the standard is not essential to the use of the test results" (Linn, 2003).

The weaknesses of standards-based statistics as summaries of performance have also been made clear. The loss of information inherent in using a coarse ordinal scale is obvious: information about changes in performance within the large bands between adjacent standards, no matter how large, go unnoted. Conversely, even very small changes in performance register as important if they happen to push students into another performance category. However, standards-based reporting has a less obvious cost as well: it can bias comparisons of trends among groups that have substantially different initial levels of performance (Koretz & Hamilton, 2006). Linn (2007) illustrated how substantial this distortion can be by examining NAEP data on the achievement gaps between minority and majority students. In grades 4 and 8, the gaps in mean scores between whites and both Hispanics and blacks narrowed between 1996 and 2005. Using the "percent above basic" and "percent above proficient" statistics, however, showed conflicting results: some comparisons showed the gap between groups decreasing, while others appeared to show an increase. This kind of distortion is a mathematical certainty: even if one constructs a hypothetical case in which all students in all groups show identical gains, standards-based measures will necessarily show the gap changing.

The response of the field to the negative evidence about performance standards fits the model of traditional psychometrics plus all too well. The operation of most large-scale testing programs has proceeded without much regard for these myriad negative findings. To be fair, the field cannot entirely resist the strong external pressure for standards-based reporting, which now currently include the statutory requirements of No Child Left Behind. There is, however, ample room for more vigorous debate about these problems and for efforts to design ways of mitigating them.

Choice of Aggregate Performance Metrics

When applied to aggregates, accountability-oriented testing requires choosing a metric to summarize performance and improvement. The measures commonly in contention often categorize schools very differently (e.g., Linn, 2000). A recent and well-publicized example is the major inconsistencies in the ratings of Florida schools between the metrics mandated by No Child Left Behind and the state's A-Plus Plan (Dillon, 2006; Peterson & West, 2006).

One aspect of this issue is the choice between value-added measures (measures that reflect the gains that students make over time) and measures that compare one cohort's performance in a given grade to the previous cohort's. The cohort-to-cohort model has dominated accountability-oriented testing in the U.S. up to the present.[1] However, interest in value-added approaches has recently become widespread, and numerous states and localities are implementing or are planning some form of value-added analysis (VAA).

Measurement issues have received short shrift in the recent debates about VAA. Some of the most salient trials of VAA have used happenstance measures of performance—whatever tests happened to be in place for other reasons, scaled however they were for those other purposes. And, the scholarly literature has focused less on measurement than on issues of statistical analysis, such as the choice of statistical model (e.g., covariate adjustment models, simple gains models, and a variety of multivariate models), problems of omitted variables, the specification of fixed versus random effects, and the impact of missing data (e.g., McCaffrey, Lockwood, Koretz, Louis, & Hamilton, 2004; Raudenbush, 2004).

Although critically important, these statistical concerns should not obscure the complex measurement issues VAA raises. The challenges are not so much new issues as extensions of traditional concerns. One example is test content. Many of the VAA approaches require tracking students over several years (e.g., for three years before a given year of evaluation). A traditional concern raised by this is that vertical linking becomes less and less tenable as the grade span increases because of content differences across the grades. However, there are additional reasons to be concerned about test content with some value-added approaches. If VAA is to be used to compare or evaluate teachers or schools, curricular differentiation becomes a serious issue, particularly in the secondary grades. For instance, in middle-school mathematics, a test that is designed for general use is likely to offer relatively weak coverage of the content that is the focus of teachers of high-track students (McCaffrey, Lock wood, Koretz, & Hamilton, 2003) and therefore may incorrectly identify these teachers as "ineffective."

A less obvious but formidable concern is posed by the problem of indeterminacy of scale. Indeterminacy refers to the fact that there is no single clearly correct way to place performance on a test onto a numerical scale, and alternative scales can provide different views of the relative performance of students or schools. The general problem of indeterminacy and its potential to undermine the robustness of inferences is well established (e.g., Spencer, 1983). Many common inferences are reasonably robust across the range of commonly used scales (Hoover, 1984), but some of the inferences entailed by VAA may be far more problematic in this respect. For example, it has been well established in the literature for more than two decades that longitudinal achievement test batteries show inconsistent trends in variance across grades. Some show the variation in student achievement increasing with grade, while others show little change, and yet others show students becoming more homogeneous. Although these disparities are often considered to be a result of the mathematical procedures used to construct the scales, the literature suggests that they result not simply from scaling, but from some unknown mix of test content, linking method, and scaling method. This indeterminacy of scale poses a serious threat to some inferences based on VAA, such as comparisons of teachers in high- and low-scoring schools (see McCaffrey et al., 2003, for discussion of these issues). Discussions of how best to address this uncertainty have barely begun.

Behavioral Issues

The behavioral issues raised by accountability-oriented testing have been explored less than non-behavioral ones, and the field has changed its practices much less in response to them. But the behavioral issues are at least as important, and they may suggest even more fundamental changes in the practice of measurement.

Discussions of behavioral responses to testing and their impact have begun to appear in theoretical discussions of the validity of inferences. In Kane's (2006) recent overview of validity theory, he provides a framework for conceptualizing these problems. In this framework, an interpretation based on performance on a test involves several stages of inference. Measurement theorists have long noted that tests are usually limited and non-random samples from the domain they are intended to represent, in part because some important aspects of performance are very difficult or impossible to assess and are therefore not included in the test design (e.g., Lindquist, 1951). In keeping with this tradition, Kane distinguishes between the target domain that the test is intended to represent and the more circumscribed subset of behaviors from which one samples in constructing a test, which he labels the universe of generalization. He calls the inference from a test score to this smaller universe of generalization "generalization" and the inference from the universe of generalization to the larger target "extrapolation" (Kane, 2006, pp. 30–33). He then cautions:

> The extent to which extrapolation remains plausible over time can depend, in part, on the stakes associated with the test. In a low-stakes context, performance in various subsets of the target domain may represent performance in the target domain fairly well. In a high-stakes context, standardization to a subset of the target domain may lead to

instruction and test-preparation activities aimed specifically at the test, thus making test performance less representative of performance in the target domain as a whole.

(p. 38)

With one important exception, Kane's discussion is consistent with a framework for evaluating validity under high stakes conditions suggested earlier by Koretz, McCaffrey, and Hamilton (2001). Koretz et al. note that the construction of a test entails not only deliberate selection of substantive elements from the domain, but also choices, some unintentional, among substantively unimportant elements, such as minor details of content, item format, style of presentation, and so forth. They note that test preparation can focus on these non-substantive elements as well as the substantively important ones, and this can substantially affect scores. This implies that behavioral responses to high stakes can threaten not only inferences from the subset of the domain eligible for testing to the larger target, but also from the specific items employed in the test to the smaller intended subdomain.

Recent empirical research has confirmed the importance of behavioral responses to accountability-oriented testing. Three outcomes bear particular note.

Distorting the Test-Taking Pool

Research in a variety of fields has demonstrated that outcomes-based accountability systems can lead practitioners to game the system by altering the pool of people from whom outcome measurements are obtained. For example, a recent study of the effects of health care report cards in New York found that they induced many cardiologists to shift certain types of care to healthier patients (Dranove, Kessler, McClellan, & Satterthwaite, 2003). Similarly, gaming of this sort has been shown in evaluations of job training programs (Heckman, Heinrich, & Smith, 2002). Numerous studies have found examples of responses of this type to accountability-oriented testing, such as increasing special education placements and retaining students in grades prior to those in which the most important tests are given (e.g., Figlio & Getzler, 2002; Jacob, 2005). Note that although these actions may bias aggregate scores by removing potentially lower-scoring students from the test-taking pool, they do not bias score-based inferences for students who are tested.

Behavioral Responses to Standards-Based Reporting

Standards-based reporting also raises behavioral issues. Educators often discuss the resulting incentives to focus disproportionate effort on the "bubble kids" thought to be near the cut scores used for accountability. In a study of the responses of educators in three states to No Child Left Behind, Hamilton et al. (2007) found that teachers reported focusing more on bubble kids. Educators express concern about this because of potentially negative effects on the students not thought to be near the cut score, but it can also bias inferences about overall school performance based on aggregate scores. Whether it biases inferences about individual students will depend on how the bubble kids are prepared for the tests.

Inappropriate Test Preparation and Score Inflation

For the science and practice of measurement, the most serious behavioral responses to testing—other than those that actually improve student learning—are those that cause score inflation, the deterioration of the relationship between the criterion and scores noted by Lindquist in the quotation above. The construction of the test requires not only sampling from the domain of knowledge and skills, but also sampling from various item formats, styles of presentation, task demands, scoring rubrics, and so forth (Koretz & Hamilton, 2006). Excessive focus on the specifics of a given test will undermine the ability of the test to represent the domain, and scores will therefore increase more than the proficiency they are intended to represent. Given the direction of the resulting bias, the term "inflation" is more descriptive than the vaguer "corruption."

Studies of validity under high-stakes conditions have compared trends in scores on a high-stakes test to trends on a lower-stakes test designed to support similar inferences about achievement (the "focal" and "audit" tests, respectively). The logic of this approach is simple: if performance on the test generalizes to the domain about which inferences are drawn, it should also generalize to scores on other tests designed to support similar inferences. As Audrey Qualls put it, to be a valid indicator of mastery of the domain, performance on the focal test must generalize to other tests with "unfamiliar particulars" (personal communication, 2004). For example, in what appears to be the first systematic study of this phenomenon, Koretz et al. (1991) wrote that:

> We expected that good performance on high-stakes tests would be caused in part by focusing undue attention on the specific content of the tests and therefore would not generalize very well when alternative measures of the same content and skills were used.
>
> (p. 2)

This is the sampling notion of validity (Kane, 1982) applied to trends over time. One would not expect the generalization to be perfect, but the more similar the targets of inference, the greater the generalization should be, and if the focal and audit tests are similar enough in terms of intended inferences, gains on the former should be substantially echoed by gains on the latter.

The few systematic studies of the validity of gains under high-stakes conditions show striking failures of generalization (Fuller, Gesicki, Kang, & Wright, 2006; Ho & Haertel, 2006; Jacob, 2005, 2007; Klein, Hamilton, McCaffrey, & Stecher, 2000; Koretz et al., 1991; Koretz & Barron, 1998; Lee, 2006; Linn & Dunbar, 1990; Schemo & Fessenden, 2003). When gains have been found on the audit test (e.g., NAEP), these have typically been one-fifth to one-third the size of gains on the focal test, and in numerous cases, rapid gains on the focal test have been accompanied by no gain on an audit test (e.g., Klein et al., 2000; Koretz et al., 1991; Koretz & Barron, 1998). Disparities this large clearly indicate that scores are inflated.

However, some discrepancy in trends may be reasonable because the focal and audit tests may not be designed to support precisely the same conclusions about achievement (Koretz & Barron, 1998; Koretz & Hamilton, 2006). For example, Reys and Lappan (2007) found variation among states in the grade-level timing of certain content standards in mathematics. Suppose that a given state's standards call for introducing a certain topic in the fifth grade, so the focal test in fourth grade does not include this topic, but the audit test does. In the absence of any score inflation, it would be reasonable to expect modestly smaller gains on the fourth-grade audit test because it includes content not taught in that grade in that state (to the extent that teachers are in fact following the timing established by the standards). Therefore, evaluating whether scores are inflated would be more difficult when the disparity in trends is smaller than in the studies cited. For the same reason, a simple difference in trends is in many cases insufficient to ascertain precisely the amount of inflation. More precise estimation of inflation would require greater specificity of inferences than one typically finds and a careful analysis of changes in performance on items tapping various aspects of content (Koretz & Hamilton, 2006).

Although differences in intended inferences may suggest that comparisons between focal and audit tests may overstate the degree of inflation, they may also understate inflation because of similarities among tests. These similarities are of two types. First, as Lindquist (1951), Kane (2006), and many others have noted, some important outcomes of education either cannot be tested with standardized achievement tests or can be tested only poorly. Therefore, to use Kane's (2006) terminology, the universe of generalization from which the tested sample is drawn is only a subset of the target about which inferences are made. Students can be inappropriately prepared for a high-stakes test in ways that make performance on the specific test, and also across the entire universe of generalization, unrepresentative of the broader target about which inferences are drawn. Because these limitations are largely unavoidable, they are shared among tests, and a comparison across tests will fail to reveal this type of inflation. Second, even within the subset of a domain that can be measured well by standardized assessments, tests are likely to show substantial similarities in content and format because of practical constraints. Therefore, the unfamiliar particulars of an audit test—to use Quall's phrase—may not be as unfamiliar as one would ideally want. These limitations would be acceptable if tests did not bear high stakes and were not used as summary judgments of entire educational programs and systems. However, when stakes are high—encouraging educators and students to focus on the particulars of the focal test—and scores from a single test are considered a summary evaluation, these limitations are serious and can lead to a substantial understatement of score inflation.

The practical importance of score inflation is exacerbated by its inconsistency. Studies have shown that inflation is highly variable across schools and over time, but they have not yet provided any information on factors that predict its severity. Furthermore, analytical methods are still insufficient to disentangle inflation from meaningful gains in scores at the level of schools or classrooms (e.g., Koretz & McCaffrey, 2005). The variability and unpredictability of inflation render comparisons among schools and classrooms suspect at best and potentially seriously misleading.

Studies of teachers' and administrators' responses to high-stakes testing, which have been an element of traditional psychometrics plus, may help explain the inflation of scores. Although the dividing line between actions that improve achievement and those that inflate scores is often hard to determine, the principle is straightforward: actions inflate scores to the extent that they boost performance on the particular test without producing generalizable gains in achievement (i.e., gains that will appear in other contexts). At one extreme, suppose that a teacher learns from the results of a large-scale assessment that his students are relatively weak in the area of ratios and proportions. If the teacher "teaches to the test" by strengthening instruction of this content, the accountability system is functioning as it should, and both scores and achievement should rise. But if the teacher's response is to look for shortcuts that take advantage of the particulars of that one test to boost scores, such as by taking time from other important aspects of the curriculum to focus on material emphasized in the accountability test, there is a risk that scores will increase more than achievement. Alternatively, some responses to testing may improve both achievement and scores. For example, some test-based accountability systems have led to a greater emphasis on writing across the curriculum (e.g., Stecher, 2002), which one would expect to improve students' writing. However, studies have also shown "rubric-driven instruction" (Stecher & Mitchell, 1995), an instructional focus on the particulars of the rubric used to score the accountability test at the expense of other important aspects of writing. This tends to inflate scores, in that improvements on the test using that one rubric will be larger than the improvement students would manifest under any other circumstances. The actions of administrators—such as establishing policies about curricula, test preparation activities, student retention, exclusion from testing, or assignment to alternate assessments; instituting "benchmark" testing; or imposing policies about the allocation of instructional time—can also boost achievement, inflate scores, or both.

We have suggested a taxonomy of test-preparation activities that helps to clarify the link between behavioral responses to testing responses and the validity of gains under high-stakes conditions (Koretz et al., 2001; Koretz & Hamilton, 2006). We use the term "test preparation" in an entirely descriptive sense to denote all activities, desirable or not, intended to improve students' performance on the test. Three types of responses to testing can be expected to produce meaningful gains in scores: teaching more, working harder, and working more effectively. These are the responses that many advocates of test-based accountability anticipate. The more interesting and difficult categories are "reallocation," "alignment," and "coaching," all of which can produce true gains in achievement, score inflation, or both.

Reallocation refers simply to the shifting of instructional resources (primarily instructional time, but other resources as well) among substantive parts of the curriculum to target better the particulars of the test. To some degree, reallocation is desirable, in that accountability tests are designed in part to signal what is important. Reallocation poses a risk, however, because tests are small and necessarily incomplete samples from the domains of achievement they are intended to represent. Allocating more time to one set of topics requires taking time away from others, and if the material that is dropped or de-emphasized is also important for the intended inferences about achievement, then scores can rise more than gains in achievement warrant. The more predictable a test is over time, the easier it is to reallocate instruction in response. Numerous surveys have found that teachers report reallocating in response to testing, including in many cases de-emphasizing important material (Stecher, 2002).

At present, alignment between tests and standards is a cornerstone of education policy. Performance standards establish what is deemed most important; tests are aligned with these to signal to teachers what is desired and to hold them accountable for producing it, and instruction is expected to change to that end. Although some degree of alignment is clearly desirable—there is nothing to be gained by testing unimportant material or by imposing tests inconsistent with the intended curriculum—alignment in itself does not protect against score inflation. Alignment is merely reallocation, but with the constraint that the material gaining emphasis in response to testing should be consistent with content standards. But score inflation depends both on what is emphasized and what is de-emphasized in response to testing and does not require a focus on unimportant material. Standards are not exhaustive, but typically focus on a subset of content goals; tests must then sample from those content areas. Therefore, alignment can entail de-emphasizing material important for inferences about achievement, thus leading to score inflation.

The final type of test preparation is coaching. This term has been used in many different ways, but we use it to denote several things: focusing on minor, unimportant details of test content; focusing on non-substantive aspects of the test, such as item format; and teaching test-taking tricks. The first of these, which we label "substantive coaching," shades into reallocation at one extreme and into cheating at the other. For example, teachers may focus on a particular way in which a type of algebraic expression is likely to appear on the test. Teaching to the rubric is an example of non-substantive coaching. Test-taking tricks include, for example, instructing students to use process of elimination rather than trying to solve multiple-choice mathematics problems. These approaches inflate scores if they generate gains that are specific to that one test and therefore do not appear in other contexts. For example, if students raise their scores on a math test by relying on a process of elimination, those gains are unlikely to appear in other contexts in which problems are not presented in multiple-choice format—for example, on certain other tests or in later employment.

Although it is obvious that score inflation undermines many interpretations of scores, it may be less obvious that it also poses difficulties for the psychometric work undergirding ongoing operational testing programs. One example is IRT non-equivalent groups anchor test (NEAT) linking, one of the methods most commonly used to link scores on accountability-oriented tests across time. In these linking methods, some secure test items (the anchor test) are reused. These anchor items serve as the constraint that allows one to make the scale comparable from one year to the next, and they provide the basis for judging the amount of overall change in achievement. The key assumption underlying this method is that if comparably proficient students perform better on these linking items in the second year, that improvement represents true gains in learning. Under high-stakes conditions, however, this assumption is not warranted. If test preparation has inflated performance on the linking items—that is, if it has improved performance on these items more than it would have improved it on other items measuring the same constructs but not sharing the same particulars—then NEAT linking will build score inflation into the scale itself. Inflation on the linking items can occur if the anchor items are sufficiently memorable that teachers can adapt their instruction to them. This can also occur even in the absence of memorable anchor items if the anchors share enough particulars with other items.

In technical terms, this entails a misplacement of the equating line. IRT NEAT linking requires that the difficulty parameters of these items be held approximately constant across years, forcing the parameter estimates for new items onto the same scale as was used the year before. In one approach, one first calibrates the data separately for each year and plots these difficulty estimates against each other. Ideally, the item difficulties in the two samples should form a straight line; items that fall far from that line are not functioning the same in the two samples and are therefore not used in linking. The position of the equating line itself, however, is treated as immaterial, and it is removed later by adjusting the item difficulty parameters in the new data. In IRT terminology, the key assumption is that the relationship between the latent trait (achievement) and observed item difficulty is fixed and that differences in estimated parameters reflected in the placement of the equating line are a function of the indeterminacy of scale that linking is designed to overcome. This is the IRT variant of the classical test theory assumption that differences in item-level performance represent differences in mastery of the underlying construct.

Behavioral responses to high-stakes testing threaten the assumption underlying NEAT linking. If inappropriate test preparation has made the linking items easier than latent proficiency warrants, the equating line will shift, and score inflation will be built into the scale for the new year of testing (Koretz & Barron, 1998; Koretz, 2007). That various factors can make an item inappropriate for linking by undermining the relationship between latent proficiency and observed difficulty is well known. For example, in a recent comprehensive review of both common practice and recent innovations in linking, Holland and Dorans (2006) mention three factors that can have this effect and thereby undermine the utility of an anchor item: changes in general knowledge, security breaches, and context effects (i.e., from anchor items being embedded in different tests in each of the two years). However, the potential impact of behavioral responses on the functioning of anchor items has as yet gained much less attention.

Although it is clear that behavioral responses to high-stakes testing pose serious challenges to conventional practices in measurement, the field's responses to them have been meager. Little has been done to explore alternative practices—either in the design of tests or in the operation of testing programs—that would be more responsive to these challenges. Perhaps most striking, the problem of score inflation gets fleeting mention, if any at all, in most evaluations or discussions of validity, whether in technical reports of testing programs, the scholarly literature, or textbooks—even though the bias introduced by score inflation can dwarf that caused by some factors that receive more attention.

TOWARD AN ACCOUNTABILITY-ORIENTED FIELD OF MEASUREMENT

Unless accountability-oriented testing becomes less important—seemingly an unlikely prospect—the model of traditional psychometrics plus will no longer be sufficient. The science and practice of measurement must begin to respond systematically to the challenges of aggregate accountability-oriented testing. Meeting this challenge will require several lines of work.

First, we need more empirical investigation of the effects of accountability-oriented testing and its implications for measurement. There is enough research to indicate that the problems are severe but not yet enough to chart adequately a future course. There are far too few studies of the validity of scores under high-stakes conditions, and we know very little about the distribution and correlates of score inflation (e.g., its variation across types of testing programs, types of schools, or types of students). Extant research on teachers' and principals' responses to testing, although somewhat more copious, is still insufficient, providing little systematic data on the use of test-preparation materials and other forms of coaching or on the relationships between test design and instructional responses. And there is too little research exploring the implications for the practice of measurement. For example, although the threat coaching poses for NEAT linking is clear, there are no empirical studies of its actual effects.

Second, we need to experiment with new approaches to test design. It is no longer sufficient to design tests with an eye to reliability and validity under low-stakes conditions. We need to design tests with an eye to improving validity under the high-stakes conditions that currently characterize much of large-scale testing. In one sense, this is not a new concern. For two decades, we have seen numerous efforts to confront this problem by creating "tests worth teaching to." In some cases, this has been taken to mean tests that comprise rich, cognitively demanding tasks; in others, it has been construed to mean tests that are aligned with standards. Both of these efforts are laudable, and both may improve the instructional effects of testing. However, neither addresses a core cause of score inflation: that is, that tests are small and incomplete samples from target domains, which in turn affords educators opportunities to take shortcuts and inflate scores.

We must work to design tests that provide less incentive to reallocate and coach inappropriately and that make those behavioral responses less damaging to validity. One particularly important question is whether tests can be designed to reduce unnecessary recurrences of details of content, styles of item presentation, construct-irrelevant task demands, and so on, and to lessen the opportunity to use test-taking tricks. In theory, introducing items with unfamiliar particulars—items that do not unnecessarily repeat details of content, that vary forms of presentation, and so on—could help provide less biased estimates of performance gains, better incentives to teachers, and perhaps even an inbuilt audit mechanism for differentiating between score inflation and meaningful gains. There is a cost to this strategy, however, as any reduction in consistency over time could threaten linking, even if only to increase random fluctuations in scores. Thus, efforts to reduce gratuitous consistency pose a tradeoff: the potential for better incentives and a reduction of bias in return for potential increases in other forms of linking error. The practicality of lessening unwanted recurrences while retaining the necessary ones remains untested, and the resource demands—in time as well as money—of attempting this could be large. Only experimentation and rigorous evaluation will clarify the degree to which innovations in test design such as this are practical and helpful.

Third, depending on the results of the evaluations noted above, the field may need to develop new methods for the procedures that maintain operational testing programs. Although the jury is still out, two of the most difficult of these may be linking and addressing the problem of scale in value-added assessments. In both cases, the problem is indeterminacy: we lack an unambiguous choice of scale, and the impact of alternative choices can be substantial. For example, most approaches to linking over time require either administering both of the tests to be linked to a single sample or holding items fixed (in the IRT context, holding their parameters constant). The former is not a realistic possibility in the case of most high-stakes testing programs, and the latter can introduce bias into the scale if the effects of coaching or other undesirable test preparation boost performance on the anchor items. This leaves us with little reason to be confident that scales are consistent and that comparisons over time are valid.

Fourth, major changes are needed in our approach to validation, that is, evaluating the degree to which score-based inferences are warranted. As long as the principal inferences based on test scores include conclusions about aggregate changes in performance achieved under high-stakes conditions, validation must include evaluating these particular inferences. This requires that we confront the problem of score inflation directly and routinely. Doing so will not be easy; it requires suitable audit measures that, in many contexts, are not available and will have to be constructed for this purpose. It will also require access to data, when it is not in the self-interest of some parties to provide it for this purpose. And, in some cases, it will require new analytical methods (e.g., Koretz & McCaffrey, 2005).

Fifth, we need routine attention to the effects of testing, regardless of whether we chose to subsume this under the label of validity. Traditional psychometrics often assumed an indirect link between testing and instruction: that the primary purpose of testing is to garner information, and that this information in turn might affect behavior. But we have long since made the transition to what Popham (1987) called "measurement-driven instruction." Tests are now intended to provide direct incentives for educators and students to change behavior, and users of scores generally assume that changes in scores on the accountability-oriented tests are themselves sufficient to determine whether those goals have been met. However, the sparse research to date has shown that this assumption is unwarranted. Therefore the quality of testing programs cannot be addressed fully without empirically evaluating impact.

This is a tremendously ambitious and as yet poorly delineated agenda. Nonetheless, the evolution of the uses of tests requires a parallel evolution in educational measurement. It is time to build on the innovative theoretical and empirical work that have characterized the current era in educational testing and to begin the development of a truly accountability-oriented science and practice of measurement.

ACKNOWLEDGMENTS

I would like to thank Michael Kane, Laura Hamilton, and Lorrie Shepard for thoughtful and constructive comments on an earlier draft of this chapter. I bear sole responsibility for any remaining errors or omissions.

NOTE

1 The No Child Left Behind accountability system is not strictly speaking a cohort-to-cohort gain model in that each school's performance in a given year is compared to a statewide target rather than to its past performance, but it is in most respects similar to a cohort-to-cohort gain approach, and in most states, no measure of individual growth is used in the accountability system.

REFERENCES

Braun, H. & Qian, J. (2007). An enhanced method for mapping state standards onto the NAEP scale. In N. J. Dorans, M. Pommerich, & P. W. Holland (Eds.), *Linking and aligning scores and scales*. New York: Springer-Verlag.

Brennan, R. L., Yin, P., & Kane, M. T. (2003). Methodology for examining the reliability of group mean difference scores. *Journal of Educational Measurement*, 40 (3), 207–230.

Burstein, L., Koretz, D., Linn, R., Baker, E., Sugrue, B., Novak, J., et al. (1995/1996). Describing performance standards: The validity of the 1992 NAEP achievement level descriptors as characterizations of mathematics performance. *Educational Assessment*, 3 (1), 9–51.

Cronbach, L. J., Linn, R. L., Brennan, R. L., & Haertel, E. H. (1997). Generalizability analysis for performance assessments of student achievement or school effectiveness. *Educational and Psychological Measurement*, 57 (3), 373–399.

Dillon, S. (2006, November 7). As 2 Bushes try to fix schools, the tools differ. *New York Times*. Online, available at: http://select.nytimes.com/search/restricted/article?res=F-60C13F C3D540C7B8EDDA00894DE404482 (accessed November 7, 2006).

Dranove, D., Kessler, D., McClellan, M., & Satterthwaite, M. (2003). Is more information better? The effects of "report cards" on health care providers. *Journal of Political Economy*, 111 (3), 555–588.

Figlio, D. N. & Getzler, L. S. (2002) *Accountability, ability and disability: Gaming the system?* (Working Paper 9370). Cambridge, MA: National Bureau of Economic Research. Online, available at: http://www.nber.org/papers/w9307 (accessed November 7, 2006).

Fuller, B., Gesicki, K., Kang, E., & Wright, J. (2006). *Is the No Child Left Behind Act working? The reliability of how states track achievement*. University of California, Berkeley: Policy Analysis for California Education.

Hamilton, L. S., Stecher, B. M., Marsh, J. A., McCombs, J. S., Robyn, A., Russell, J. L., et al. (2007). *Standards-based accountability under No Child Left Behind* (MG589). Santa Monica: RAND. Online, available at: http://www.rand.org/pubs/monographs/2007/RAND_MG589.pdf (accessed June 26, 2007).

Heckman, J. J., Heinrich, C., & Smith, J. (2002). The performance of performance standards. *Journal of Human Resources*, 37 (4), 778–811.

Ho, A. D. & Haertel, E. H. (2006). *Metric-free measures of test score trends and gaps with policy-relevant examples*. CSE Technical Report 665. Los Angeles: Center for the Study of Evaluation, University of California.

Holland, P. W. & Dorans, N. J. (2006). Linking and equating. In R. L. Brennan (Ed.), *Educational measurement* (4th ed., pp. 187–220). Westport, CT: American Council on Education/Praeger.

Hoover, H. D. (1984). The most appropriate scores for measuring educational development in the elementary schools: GEs. *Educational Measurement: Issues and Practice*, 3 (4), 8–14.

Jacob, B. A. (2005). Accountability, incentives and behavior: The impact of high-stakes testing in the Chicago public schools. *Journal of Public Economics*, 89 (5–6), 761–796.

Jacob, B. A. (2007). *Test-based accountability and student achievement: An investigation of differential performance on NAEP and state assessments* (Working Paper 12817). Cambridge, MA: National Bureau of Economic Research. Online, available at: http://www.nber.org/papers/ w12817 (accessed November 7, 2006).

Jaeger, R. M. (1989). Certification of student competence. In R. L. Linn (Ed.), *Educational measurement* (3rd ed., pp. 485–514). New York: American Council on Education/ MacMillan.

Kane, M. T. (1982). A sampling model for validity. *Applied Psychological Measurement*, 6 (2), 125–160.

Kane, M. T. (2006). Validation. In R. L. Brennan (Ed.), *Educational measurement* (4th ed., pp. 17–64). Westport, CT: American Council on Education/Praeger.

Kane, M. T. & Brennan, R. L. (1977). The generalizability of class means. *Review of Educational Research*, 47 (1), 267–292.

Kane, T. J. & Staiger, D. O. (2002). The promise and pitfalls of using imprecise school accountability measures. *Journal of Economic Perspectives*, 16 (4), 91–114.

Klein, S. P., Hamilton, L. S., McCaffrey, D. F., & Stecher, B. M. (2000). *What do test scores in Texas tell us?* (Issue Paper IP-202). Santa Monica, CA: RAND. Online, available at: http://www.rand.org/publications/IP/IP202/ (accessed January 12, 2004).

Koretz, D. (2007). Using aggregate-level linkages for estimation and validation: Comments on Thissen & Braun & Qian. In N. J. Dorans, M. Pommerich, & P. W. Holland (Eds.), *Linking and aligning scores and scales* (pp. 339–353). New York: Springer-Verlag.

Koretz, D. & Barron, S. I. (1998). *The validity of gains on the Kentucky Instructional Results Information System (KIRIS)* (MR-1014-EDU). Santa Monica: RAND.

Koretz, D. & Hamilton, L. S. (2006). Testing for accountability in K–12. In R. L. Brennan (Ed.), *Educational measurement* (4th ed., pp. 531–578). Westport, CT: American Council on Education/Praeger.

Koretz, D. & McCaffrey, D. (2005). *Using IRT DIF methods to evaluate the validity of score gain* (CSE Technical Report 660). Los Angeles: Center for the Study of Evaluation, University of California.

Koretz, D., McCaffrey, D., & Hamilton, L. (2001). *Toward a framework for validating gains under high-stakes conditions* (CSE Technical Report 551). Los Angeles: Center for the Study of Evaluation, University of California.

Koretz, D., Linn, R. L., Dunbar, S. B., & Shepard, L. A. (1991, April). The effects of high-stakes testing: Preliminary evidence about generalization across tests. In R. L. Linn (Chair), *The effects of high stakes testing*. Symposium conducted at the annual meetings of the American Educational Research Association and the National Council on Measurement in Education, Chicago.

Lee, J. (2006). *Tracking achievement gaps and assessing the impact of NCLB on the gaps: An in-depth look into national and state reading and math outcome trends.* Cambridge, MA: Civil Rights Project at Harvard University.

Lindquist, E. F. (1951). Preliminary considerations in objective test construction. In E. F. Linquist (Ed.), *Educational measurement* (2nd ed., pp. 119–158). Washington: American Council on Education.

Linn, R. L. (1997). Evaluating the validity of assessments: The consequences of test use. *Educational Measurement*, 16 (2), 14–16.

Linn, R. L. (2000). Assessments and accountability. *Educational Researcher*, 29 (2), 4–16.

Linn, R. L. (2003). Performance standards: Utility for different uses of assessments. *Education Policy Analysis Archives*, 11 (31). Online, available at: http://epaa.asu.edu/epaa/v11n31/ (accessed October 20, 2003).

Linn, R. L. (2007, January). Educational accountability systems. Paper presented at the meeting of the National Center for Research on Evaluation, Standards, and Student Testing, Los Angeles.

Linn, R. L. & Dunbar, S. B. (1990). The nation's report card goes home: Good news and bad about trends in achievement. *Phi Delta Kappan*, 72 (2), October, 127–133.

Linn, R. L., Baker, E. L., & Dunbar, S. B. (1991). Complex, performance-based assessment: Expectations and validation criteria. *Educational Measurement: Issues and Practice*, 20 (8), 15–21.

McCaffrey, D. F., Lockwood, J. R., Koretz, D. M., & Hamilton, L. S. (2003). *Evaluating value-added models for teacher accountability* (MG-158-EDU). Santa Monica: RAND.

McCaffrey, D. F., Lockwood, J. R., Koretz, D., Louis, T. A., & Hamilton, L. (2004). Models for value-added modeling of teacher effects. *Journal of Educational and Behavioral Statistics*, 29 (1), 67–101.

Messick, S. (1989). Validity. In R. L. Linn (Ed.), *Educational measurement* (3rd ed., pp. 13–103). New York: American Council on Education/Macmillan.

Peterson, P. E. & West, M. R. (2006). Is your child's school effective? Don't rely on NCLB to tell you. *Education Next*, (6) 4, 76–80.

Popham, W. J. (1987). The merits of measurement-driven instruction. *Phi Delta Kappan*, 68 (9), 679–682.

Popham, W. J. (1997). Consequential validity: Right concern—wrong concept. *Educational Measurement: Issues and Practice*, 16 (2), 9–13.

Raudenbush, S. (2004). Schooling, statistics, and poverty: Can we measure school improvement? The ninth annual William H. Angoff Memorial Lecture. Princeton, NJ: Educational Testing Service.

Reys, B. & Lappan, G. (2007). Consensus or confusion: The intended math curriculum in state-level standards. *Phi Delta Kappan*, 88 (9), 676–680.

Schemo, D. J. & Fessenden, F. (2003, December 3). Gains in Houston schools: How real are they? *New York Times*, pp. A1, A27.

Shepard, L. A. (1994). Implications for standard setting of the National Academy of Education evaluation of the National Assessment of Educational Progress Achievement Levels. In *Proceeding of the joint conference on standard setting for large-scale assessments, Vol. II* (pp. 143–160). Washington, DC: National Assessment Governing Board and National Center for Education Statistics.

Shepard, L. A. (1997). The centrality of test use and consequences for test validity. *Educational Measurement: Issues and Practice*, 16 (2), 5–8, 13, 24.

Spencer, B. D. (1983). On interpreting test scores as social indicators: Statistical considerations. *Journal of Educational Measurement*, 20 (4), 317–333.

Stecher, B. M. (2002). Consequences of large-scale, high-stakes testing on school and classroom practices. In L. S. Hamilton, B. M. Stecher, & S. P. Klein (Eds.), *Making sense of test-based accountability in education* (pp. 79–100). Santa Monica, CA: RAND.

Stecher, B. M. & Mitchell, K. J. (1995). *Portfolio-driven reform: Vermont teachers' understanding of mathematical problem solving and related changes in classroom practice* (CSE Technical Report No. 400). Los Angeles: Center for the Study of Evaluation, University of California.

HOW DO ACCOUNTABILITY, ASSESSMENT, EVALUATION, AND RESEARCH APPLY TO MY PROFESSION?

As teachers and educational leaders, you may feel that your voice is not heard under the current landscape of accountability. Although those feelings are understandable, please know that you do hold a lot of power. Your actions and decisions have significant consequences on the lives of students and other educators. You also have the power to shift policy. Do not forget that you are educational experts and you can be the voice for your professions. You do have a voice and you have much control over what happens within the walls of your classrooms and/or your buildings. There is no better way to make your voice heard than by using appropriate assessments that produce reliable data by which you can make appropriate conclusions or inferences about students, teachers, curriculum, and various programs that you've instituted within your schools. Thus, knowing how to differentiate between each of the described concepts is fundamental to ensuring decisions and conclusions are defendable to both internal and external stakeholders.

CHAPTER 3

Classical Test Theory

AS YOU READ

As you read the chapter, think about the following:

- How do **you** define or explain classical test theory?
- What is **your** experience with classical test theory?

WHAT ARE THE GENERAL PRINCIPLES OF CLASSICAL TEST THEORY?

There are multiple test theories; some more prevalent than others in the United States. **classical test theory (CTT)**, sometimes referred to as *true score model*, has been around for more than a century and it is fairly accepted. There are, however, several critics as well. The simplistic and basic premise of the theory is that measurement error, a random latent variable, is always part of an observed score. This means that no matter the quality of the assessments, there will always be error. There are various types of error in measurement. For example, both random and systematic error affect assessment results. **Random error**, sometimes called internal error, refers to effects that impact the results of the particular assessment for specific individuals at that time. Let's think about a typical classroom example and review how random error can impact those results. As a middle-level civics teacher, you give each of your 75 sixth-grade civics students an assessment of their knowledge on the three branches of government. Each student had a different experience the night before. Ten of those students were up all night for various reasons. One of the students woke up to arguing parents. Another woke up with a headache. And another was in an argument with her best friend just before the test. Three other students just found out that they achieved honor roll for that particular grading period. Another was just informed that she would receive recognition for a recent accomplishment. In all instances, these students complete the exam with these situations on their minds. You, as the teacher, grade the exams and realize that some of those students did not perform as well as they typically do. Others performed higher than typical. So, you talk to them to find out their explanation for their performance on the exam. This scenario is fairly probable and common in everyday classroom settings. These situations are all examples of random error. Random error is caused by various factors that randomly affect scores on an assessment or in technical terms—randomly affect variables across the sample (your students). Random error does not consistently impact the results of the entire sample as some scored lower and others scored higher than typical. Random error is often referred to as "noise" in the data.

Systematic error, sometimes called external error, differs from random error in that it affects the entire sample. For example, you administer the same exam as described above. However, construction workers are jackhammering the concrete just outside the classroom window the particular day that you administer the exam. The noise is loud and those inside the classroom walls feel the vibration throughout the day. This noise is likely to cause slightly lower scores of all examinees. Another example, and possibly the most important, is error embedded within the assessment. If the items/questions on the assessment tool itself are written poorly or there are other structural issues with the assessment instrument itself, the scores of all examinees may be affected. In these cases, the error is considered systematic because it affects the entire sample. Systematic error if often referred to as "bias" in the data.

You must know that there is always error when you try to measure variables—regardless of whether those variables are latent or observable. Let's go back to our example above that includes our desire to assess the students' knowledge of a concept. How do we measure what students know? As stated above, we can use an existing or develop a new test to assess their knowledge. So, as a teacher, you decide to use an existing assessment instrument to measure the students' knowledge about the galaxy. You determine that this is a good instrument to use and thus the results are equal to the students' knowledge. Not so fast—can we actually measure what an individual knows? Actually, we cannot but we use students' responses to questions on instruments to provide an estimate of what they actually know.

There is no way for us to directly measure what the student knows because knowledge is a **latent construct**, which is a variable that is not directly measurable or observable. Some examples include honesty, knowledge, thoughts, beliefs, trustworthiness, happiness, and so on. Measuring latent variables inherently include error and conclusions are inferred based on a mathematical model. Now, think about measuring something that you can directly observe.

Observed variables are things that you can directly measure using scales. Some examples include weight, pH levels, speed, time, temperature, etc. Although these are directly observable, there is still error when you try to measure them. Often, you do not make conclusions based on one measurement. If you think about the simple concept of your weight. Let's say that you wake up in the morning and step on to a scale to determine your weight. You step on that scale and the number appears. You think to yourself, "Wow, I must have gained a lot of weight based on this number. I wonder if that is accurate." So, you get off the scale and then step on to the scale once again to ensure the accuracy of the number. This time, you get a lower number. You then think to yourself that this number appears more accurate, but you aren't positive which is correct. So, for a third time, you step onto the scale. Now, this time, the number is the same as the second reading of your weight. Thus, you conclude that this must be a more accurate representation of your weight. The point is, even when you measure variables that are considered directly observable, there is error in the measurement. This is the fundamental principle of classical test theory.

Now, let's apply CTT to assessment and measurement to a simple classroom example. As educators, we want to determine what our students know and what they do not know. So, we measure students' knowledge by creating or selecting questions to assess their knowledge. Then, we determine an appropriate numerical scale and the properties of that scale. Let's say that for this test/assessment of knowledge, we use a 100-point scale. There are 25 questions on the test and each question is worth four points. The responses are either marked correct or incorrect (no partial credit). Student 1 receives 50% or 50 points on the test. Student 2 receives 100% or 100 points on the test. Student 3 receives 0% or 0 points on the test. Does Student 2 know twice as much as Student 1? Does Student 3 know absolutely nothing about the concept? These are false conclusions that are sometimes made in the classroom. The correct conclusion is that we created the scale and the properties of the scale to determine a grade. We hope that the grade is a reflection of what the students know; however, we cannot make the conclusion or assumption that Student 2 knows 100% of the information simply because he received 100% on the assessment. We cannot assume or conclude that Student 3 knows nothing about the concept simply because he received 0 on the test. And we certainly cannot conclude that Student 2 knows twice as much as Student 1 simply because he scored twice as high.

The concept of a **true score** is often difficult to grasp. I believe this is probably the case because a true score is an abstract concept that is impossible to obtain in an educational setting. A true score is the real score of something (knowledge, skills, attitude, height, etc.). As mentioned above, it is not possible to obtain a true score in the real world. Even in the case of measuring a person's height—there is measurement error involved. So, if you really wanted to determine a true score for some measure (e.g., height), then you would have to collect data an infinite number of times. Again, this is impossible in the real world: The true score is essentially the average score of something measured an infinite number of times. So, to determine a child's knowledge about the sun, first you'd have to develop an assessment or multiple parallel assessments that elicit both reliable and valid results. Assuming there is no error, you'd give the assessment to the child an infinite number of times. Then, average those final scores of the infinitely given error-free assessments and voilà—you've obtained a true score. Now, let's get back to the real world. It is extremely important to ensure that the scores from assessments have strong evidence of both reliability and validity because typically, in a classroom setting, we only assess students one time. Then, we make decisions based on the results of the students taking the test one time (students' grades, retention, free time, punishments, rewards, etc.). It is highly unlikely that the students' observed scores are equivalent to their true scores. Please know that you are never actually collecting a student's true score when you give an assessment of any kind. There is always error involved. Your job is to ensure there is as little error as possible and that the scores are as reliable as possible for you to make valid conclusions based on those scores.

CTT assumes that each examinee has an innate true score but measurement error is central to the theory. Thus, the CTT equation is the following:

$$X = T + E$$

Where

X represents the observed score
T represents the true score
E represents the error

The **standard error of measurement (SEM)** is an estimate of how repeated measures of the same student on the same test are distributed around that student's true score. In simple terms, it represents the accuracy a single score approximates the true score. Remember, we can't actually measure a true score so we calculate the SEM to determine how much error is apparent. SEM is directly related to the reliability meaning the larger the SEM, the lower the reliability. The equation for calculating the SEM is represented as follows:

$$S_E = S_X\sqrt{1-\alpha}$$

Where:

S_X represents the sample standard deviation of test scores
α represents alpha, the estimated reliability coefficient

In a later chapter, we will discuss reliability in detail. But, for one to fully understand CTT, one must have a conceptual understanding of reliability. In its most general sense, the **reliability coefficient** refers to the dependability or consistency of scores. For example, if you were to give students the same test multiple times (assuming they did not remember any of the questions) and they scored widely different on each administration, then the results would not be consistent or dependable. Conversely, if you administered a test multiple times to the same students and the results were very similar across administrations, then the results would be consistent or dependable. The number of items on the test has a tremendous impact on the reliability coefficient. This is the basic premise of the concept of reliability. We will learn about how to determine estimates of reliability in a later chapter.

Confidence intervals (CIs) are a range of scores that have a specified probability of including the student's true score. In education, CIs are often referred to as score bands. Because we know we cannot measure a true score, we calculate a range of scores for which we can be confident or the probability that the student's true score will fall within. Typically, we use 68%, 95%, or 99% CIs. So, let's say that a student received a raw score of 85 (out of 100) on a classroom test. After calculating a CI for that score, we could state, "There is a 95% chance that the student's true score for that test is between xx and xx." Or another way to state it, "Given the student's score of 85, there are nine and a half out of ten chances that the student's true score will fall between xxx and xxx." To actually calculate the CI, you simply add and subtract the SEM (one, two, or three SEMs depending on the level of confidence you think is appropriate 68%, 95%, or 99%) from the raw score—this will give you the range or the CI.

Score +/– 1 SEM (68% CI)
Score +/– 2 SEM (95% CI)
Score +/– 3 SEM (99% CI)

CIs can also be calculated for a group of scores. Typically, if you are calculating CIs for a group of scores; then you would add and subtract the standard deviation from the mean of scores. Remember, if you want to determine the CI for one student's score, then you add and subtract the SEM from the student's observed score. If you want to calculate the CI for a group of students, then you add and subtract the standard deviation from the mean score of the group.

WHAT DOES THE LITERATURE SAY ABOUT CLASSICAL TEST THEORY AND STANDARD ERROR OF MEASUREMENT?

Reading 3.1

CLASSICAL TEST THEORY AND HIGHER EDUCATION: FIVE QUESTIONS

Elizabeth Schaughency,[1] Jeffrey K. Smith, Jacques van der Meer, and David Berg

It is not too difficult to measure the height of a child. But how we do measure the height of her proficiency in reading? Or how well she has mastered the first half of European History 252, or determine if her academic skills would predict that she would be successful at our university?

Classical Test Theory (CTT) refers to a body of psychometric theory about how to understand how well we have measured some mental characteristic or trait of an individual. How do we know that we have obtained a measure of unobservable qualities of a person such as knowledge, skills, or other psychological correlates of academic success on which we can rely? What would such a question even mean, exactly? These phenomena cannot be measured directly.

To be a bit more technical, CTT typically refers to issues of reliability and measurement error. As such, CTT may be considered to be a psychometric consideration, concerned with the development and refinement of educational and psychological measurement. For sake of completeness, we discuss issues of validity in this chapter as well. This is not meant to be a technical treatise on these issues, as excellent examples of those exist elsewhere in this volume; rather we want to focus on the fundamental ideas of CTT and contrast them to modern theories and approaches to measurement development, such as item response theory (IRT) and evaluating technical adequacy of approaches designed to assess learning over time or in response to instructional intervention.

CTT (which wasn't referred to as CTT until it was essentially superseded by IRT) had its statistical roots in the turn of the last century with the development of the correlation coefficient, and with most of the formulae that are used today being derived mid-20th century (Cronbach, 1951; Guttman, 1945; Hoyt, 1941; Kuder & Richardson, 1937). Perhaps the most comprehensive explication of the theories that underlie classical test theory can be found in Lord and Novick (1968). Interestingly, they also presented the fundamentals of IRT theory, which has largely replaced CTT in big test development projects, such as development of new, published norm-referenced tests. This raises the question of "Why look at CTT at all if we have a new and more sophisticated approach?" The answer to this question is that CTT works quite well for many tests developed for more limited, or local, use, and still provides useful descriptions of the consistency of measurement of assessment instruments used today.

In this chapter, we seek to do several things. First, we want to look at what kinds of testing/assessment goes on in higher education today, and how the ideas of CTT potentially contribute to those activities. Second, we want to examine how classical test theory works from a conceptual basis—what are the essential ideas behind CTT. Third, we will consider how CTT differs from IRT, especially as it applies to higher education. And finally, we will discuss how CTT impacts on measurement issues in higher education today, and the need to be mindful of the purpose of assessment when considering how best to evaluate whether assessments provide good information for educational decision making.

WHAT DO WE TEST IN HIGHER EDUCATION?

Testing serves a wide range of functions in higher education. The influence of testing in higher education begins well before students unpack their bags in their first dorm room as the first function of testing in higher education may include selection of students into universities (see Zwick [...]). For example, in the US, this is done through what are probably the two most famous tests in American education: the SATs, and the ACTs. The SATs have been in existence in one form or another since 1926, and a precursor to it since 1901 (Donlon, 1984).

College entrance examinations serve several potential functions in the higher education admissions process. First, they give students an idea of what kinds of postsecondary institutions are likely to admit them. Students can compare their performance to the typical profile of the student body at a college or university. Next, the examinations give the institutions an independent indication of the achievement/ability level of a given student.

A second area of assessment that is common in higher education is skills testing for incoming students (see Morgan [...]). Most American two-year colleges and many four-year colleges and universities assess the fundamental skills of students to see who would benefit from instruction in basic skill areas. Related to this type of testing is testing that occurs to place students in the proper levels of regular course sequences, for example, in which level of French should a student be enrolled?

A third area of testing is perhaps the one most familiar to higher education faculty members, and that is the testing that goes on as a natural part of instruction. We test students in courses in order to determine mastery of course content, usually so that grades can be assigned in courses (see Drezek McConnell & Doolittle [...]). This testing may also involve assessments that are used to help students and faculty determine how well students are progressing through the course material, and what the best instructional steps to take next are. Testing of academic achievement used for grading is an example of summative assessment, and testing to determine how to tailor instruction to meet students' needs is an example of formative evaluation. Most testing that goes on at the college level is summative in nature, but faculty would probably be well advised to consider formative assessment in their courses to enhance learning and instruction.

Oftentimes, we use alternatives to testing in college and university courses, opting instead for students to write papers, make presentations, and so forth. If these are used as part of a grading program, then many of the ideas presented here will apply in theory (if not in actual application) to those types of assessments.

WHAT ARE THE CHARACTERISTICS OF GOOD MEASUREMENT?

Tests are administered to serve a purpose. As can be seen in the section above, tests that are used in higher education can serve different purposes. We might want to use a test in order to pick the best students for our university, or to determine which ones need extra help to be successful, or to determine what grade a student should receive in a course. Since the purposes for tests can be different, to a degree, the evidence we consider in evaluating a test may differ as well (see chapters by Geisinger, Shaw, & McCormick; Pitoniak & Morgan [...]). But the general ideas of what to consider when selecting a test don't change. What we want in a test is something that helps us to make good educational decisions. That decision might be to admit or reject an applicant from our university, to determine that a person should take a remedial mathematics course or not, or whether a person has earned an A or a B in our course. These are all decisions that are augmented by having test information.

In each of these cases, we want information that contributes to a good decision. We might even say "the correct decision," or "the best decision." So what should we consider in order to conclude that assessment findings have fulfilled that goal? Historically, we say that we should consider evidence supporting reliability and validity for given purpose (AERA, APA, & NCME, 1999).

Validity refers to the degree that the decisions that we make based on test information are valid ones, or ones that are sound and justifiable. Reliability is a somewhat narrower concept; the idea here is that a test should provide information that is dependable. For example, assessment results should be consistent over time if the skills and knowledge being measured have not changed for the individual. Since there are other chapters on both validity and reliability in this volume, and since CTT is primarily focused on reliability, reliability takes the lead in this discussion, but we will look at issues of validity as well, in particular as they impact on the ideas of reliability—and vice versa.

THE BASIC CONCEPTS OF CTT: WHAT DO TRUE SCORE, THE STANDARD ERROR OF MEASUREMENT, RELIABILITY, AND VALIDITY REALLY MEAN?

Suppose we are interested in assessing vocabulary knowledge. Vocabulary is a convenient example for a number of purposes. To begin, on the face of it, we have a notion of a definable universe of possible test items. And we can realistically think of taking a random sample of items from that universe. As noted above, CTT is largely concerned with evaluating how good the data that we obtain from test results are. A basic assumption of CTT is that test scores are the sum of two theoretical components, the true score and error (Christ & Hintze, 2007). Therefore, in order to understand the basic concepts that underlie CTT, it is probably best to start with the notion of the true score.

What Is a True Score?

Although the concept of a true score has several definitions with fine philosophical differences (Lord & Novick, 1968), we will overlook the nuances in favor of a simpler, yet accurate definition. The true score is the average score a person would get if he/she took the same test over and over again, with no opportunity for learning or skill development and all memory of having done so being erased in between administrations. Continuing with our example, suppose we wanted to know what percentage of all the words in the English language a person knew. We could randomly draw a sample of 100 words from the dictionary and ask a person to give definitions of them. We could mark them as right or wrong, and then calculate a percentage of words defined correctly. We could then sample another 100 words and repeat the process, and then do this again and again. The average percent correct after an infinite number of repetitions would be the person's true score. Or we might simply give the person all the words in the dictionary at one time. We would get the same estimate either way.

What Is Standard Error of Measurement?

Having defined a true score, we can get an idea of what a standard error of measurement is. We can think of the fact that some samples of items from the dictionary will be easier than others, and we can think that some samples might "fit" better with testees than others. Thus, there will be some variability in performance depending on the words that are sampled in the test. Additionally, our testee may be feeling better on one day than the next, or have something troubling on his mind that interferes with his performance on a given day. Therefore, there might be variability due to the testee as well as due to the set of items. In sum, CTT thinks of a test score as having two components, a true score which represents the true ability of the person, and an error score that represents the vicissitudes of the particular test and of the individual on the day of the test. Formally, we say that an observed score (X) on a test is equal to the true score (T) of the individual and an error score (E) that is peculiar to the particular administration of that test to that person on that day, or $X = T + E$.

Harkening back to our poor soul who took an infinite number of vocabulary tests, we could not only calculate a mean for that person, but indices of variability in performance around that mean as well. For example, we could calculate the standard deviation of the distribution of scores generated by giving this person an infinite number of vocabulary tests. Knowing the standard deviation would give us a very good (probabilistic) idea of how far from the true "performance" of the person any given test result might be. This provides the conceptual definition of the standard error of measurement. And it makes good sense. If we see a person get a 61% correct on the vocabulary test, and we calculate the standard error to be 4%, then, by assuming (reasonably) a normal distribution of scores (the central limit theorem comes into play here), we could say that the person's true score would be likely (68% of the time) to be between 57% and 65% (plus and minus one standard error), and be very likely (96% of the time) to be between 53% and 69%. If this level of accuracy does not seem sufficient, then we need a test with a smaller standard error of measurement (perhaps by increasing our sampling of items).

So that takes care of true score and standard error of measurement. How about reliability and validity? We've taken an initial shot at definitions above, but let us revisit and elaborate a bit on those and discuss how they relate to the concepts just discussed.

What is Reliability?

Accuracy and reliability of measurement are related, but not synonymous (Christ & Hintze, 2007). However, in general, the less evidence we have to support reliability (consistency, dependability) of measurement, the less confident that we can be that test results provide a trustworthy estimate of the student's true performance. There are a variety of ways to assess reliability, as described by Thompson and Vacha-Haase. We outline some of these below, presenting the ideas as they developed. As will be seen, a common thread across these approaches is to examine the degree to which two samples of performance correspond, or correlate. As will also be seen, however, because these reliability indices are based on types of different information, they convey different information—something to be kept in mind when selecting measures for educational decision making.

Consistency in Performance Across Time

To begin, imagine that our vocabulary test was given to a sample of individuals, and then the following week was given again. A correlation is calculated between the two administrations, and that correlation is the estimate of the reliability coefficient. It gives us an index of how consistent scores will be from one administration to the next. It is what is called *test-retest reliability.* From a test selection perspective, the most relevant test-retest reliability periods are those that parallel the decision-making period (Macmann & Barnett, 1999). For example, if we are to use our measure of vocabulary knowledge as an indication of the student's vocabulary at entry into postsecondary education and our vocabulary measure was typically administered during the senior year of high school, ideally we would want evidence of test-retest reliability that spans that period.

But what if people become curious about the items on the test that they were not sure about? What if they found a number of the definitions? And to further complicate matters, what if some of the testees do this and others do not, how would this affect our interpretation of assessment results? The problem here is that we have what is called measurement reactivity, that is, the process of being assessed affected the test-takers' assessments in some way. It is not as severe as the Heisenberg Principle of Uncertainty in quantum physics that acknowledges a trade-off in the precision with which we can measure some pairs of properties (the more precisely we measure position, for example, the less precisely we measure momentum), but it does hold potential for disturbing our estimate of the reliability in that those participants who took the test at Time 2 may differ from when they took the test at Time 1 due to their experience with test at Time 1.

To solve this problem, the notion of a parallel form of a test was invented. A parallel form of a test has the same number of items with roughly the same mean, standard deviation, and so forth. It is as close as possible to the original test, with all new items. This is not hard to conceptualize with vocabulary items, but is a bit tougher with introductory psychology. Now, instead of giving the same test twice, the second test is what we hope to be a parallel form of the test. We again correlate the results of the two administrations, and the correlation becomes our estimate of the reliability. It is called *parallel forms reliability.* It solves some of the problems of measurement reactivity, but not all of them. Thinking of the biology test, some testees may still decide to investigate certain areas that intrigued them on the test, and thus are not really the same people that they were (with regard to introductory biology) when they first took the test. As may be obvious, parallel forms reliability becomes an important consideration when assessments are to be conducted on more than one occasion, as in program evaluations that seek to evaluate the effectiveness of instructional strategies through the administration of pre- and post-testing to participants.

Both test-retest and parallel forms reliability, however, typically assess the stability of the measure over a short period of time, a period where we would expect there to be no change in the abilities being measured. There are additional practical problems with both test-retest and parallel forms reliability. Few people really want to come back a week later and take a test again if they don't have to.

Consistency Across Content

The question arises as to whether one could have testees take the test and the parallel form of the test in the same sitting, thus obviating the need for a second session? Well, if we could imagine a test twice as long as the test in which we were interested, then we could divide the results in half, correlate the two halves, and we would have our estimate of parallel forms reliability without the problem of measurement reactivity or the inconvenience of two sessions. We could, perhaps, make one form of the test be all the odd numbered items, and the other half be all the even numbered items. We then add up the odd items and then the even items, and correlate the two totals.

The only problem here is that this requires people to sit for a rather long test. To solve this problem, measurement specialists give a test of the length in which they are interested. They then correlate a test half the length in which they are interested with the other half of the test. This is called a *split-half estimate* of reliability. But there is a problem here. This is an estimate of a test half the length of the real test, and it only makes sense that a longer test would, ceteris paribus, be a more reliable test. So an adjustment formula is applied that takes the correlations from half the test and then adjusts that correlation to what it would be for full-length tests. This is done with what is called the Spearman-Brown Prophecy Formula. For an excellent and detailed explanation of this formula and all of reliability theory, see Feldt and Brennan (1989).

A test could be split in half in any of a large number of ways, not just odd items and even items. A formula was developed that essentially takes all the possible ways of splitting a test in half and estimating the reliability, and then averages all those possibilities. It is called Kuder-Richardson Formula 20 (yes, actually the twentieth formula they came up with). This formula, known as KR20, is widely used still in testing. It is based on the notion that all the items will be scored in a right/wrong fashion. Cronbach (1951) came up with a formula that generalizes KR20 to the case where the items can be scored on scales instead of just right/wrong. We call this coefficient alpha, or sometimes, Cronbach's alpha.

The split-half reliability estimate, KR20, and coefficient alpha do not look at the stability of measures over time as they all use data from a single administration of a test. Instead they provide indices of the extent to which items comprising a test seem to be tapping into a similar domain and are referred to as *internal consistency* measures of reliability. There are a variety of technical issues in the development of these concepts and the estimating algorithms of the concepts and coefficients that are not our concern here as they are described well elsewhere in this volume.

Consistency Across Observers

There is a final class or category of reliability coefficients that are typically associated with the classical model. These began by considering the issue that some tests do not have simple right/wrong answers such as multiple choice items, but require scoring that is based on the judgment of scorers or raters. We know that not everyone agrees on everything all the time, so there will be some variability in scores due to differences in raters. Inter-rater reliability addresses the issue of variability of raters.

Work in the area of inter-rater reliability led to the development of generalizability theory (Brennan & Kane, 1979; see also Webb, Shavelson, & Steedle [...]). Generalizability theory provides the statistical basis for looking at variability in test performance from any possible definable source (persons, items, raters, occasions, etc.). In sum, although the various reliability indices all speak to the issue of dependability in measurement, they consider different potential sources of information in doing so. In practice, we should consider types of reliability evidence that are most relevant to our educational decisions. For example, if we are making decisions based on how someone graded a test, we are interested in evidence of inter-rater reliability. If we are interested in making decisions about a student's likely performance over time, then we are interested in evidence of test-retest reliability.

What Is Validity?

We selected assessment of vocabulary as a convenient illustrative example. With vocabulary, we seem to have a clearly defined skill set that we are measuring. Vocabulary knowledge seems a straightforward notion. If we are trying to measure knowledge of introductory biology, or readiness to take English 101, or who are the best students for our university, the notions of what it is that we are trying to assess become a bit murkier. Different teachers might not agree even on what introductory biology should consist of, much less how to generate items to measure it.

The theoretical description of the concept that we want to measure is referred to as a "construct." We use the term *construct* to acknowledge that the phenomenon that we are measuring is inferred and that we are not measuring it directly but only indirectly through our assessment (see G. T. Smith, 2005, for discussion). What is the construct to be measured in introductory biology or in who are the best students for our university? The issues related to answering those questions, and in determining whether we have done a good job of answering them fall under what is known as *construct validity* (Linn, 1994).

In general, validity evidence speaks to the degree to which our score is measuring what we want it to measure. Is it giving us the information that we are looking for? Will it lead us to make good decisions in our situation and for our purpose? Often, we hear or read about the concurrent validity, or predictive validity, or face validity of a test. But really this should be considered to be shorthand terminology for saying that we have evidence of the validity that is classified as concurrent in nature (i.e., providing information regarding the extent to which results correspond to other indicators of the construct at the same time), or predictive (providing information regarding the extent to which results predict performance at a future time), and so forth. All validity evidence essentially speaks to the same question: are we getting information that will lead to good decisions? Validity evidence accumulates as studies are conducted that speak to the issue of validity with given samples or for given purposes (AERA, APA, & NCME, 1999).

We should also consider the notions of incremental, treatment or instructional, and consequential validity evidence. *Incremental* validity evidence refers to the extent which the assessment results contribute new information and add to our decision making (Hunsley & Meyer, 2003). For example, is there value added in asking students to take a pre-enrolment placement exam or would we make the same advising decision based on available high school records?

Treatment or *instructional validity* again refers to contributions to educational decision making but asks a different question: Rather than asking whether the test predicts student performance, the question is whether assessment results inform professional decision making (L. S. Fuchs, 2004). Learning from assessment results that a student may be academically under-prepared for university study is different than learning what that student's instructional needs are and how best to support them. Instructional validity, sometimes referred to as instructional utility, refers to this latter issue. Finally, perhaps ultimately, *consequential validity* should be considered. Consequential validity considers the question of the consequences of using a particular measure as opposed to using an alternative approach, or none at all (AERA, APA, & NCME, 1999). Although we will not consider validity issues in detail here, we will return to these ideas as we explore the type of testing done in higher education.

Concluding Comments on Basic Concepts

One final important point to emphasize is that the information that is obtained on the quality of a test is only appropriately used for people who are assessed with the same form of that particular test and who are similar to those for whom the reliability estimates were obtained (and validity for that matter). That is to say, the estimates are *sample-specific*. This is more important than it might initially seem. A test that has reliability and validity evidence for predicting whether applicants who are English language learners are likely to be successful in their beginning university studies may not predict equally well for students whose first language is English. Similarly, a test that predicts grades in some university coursework may or may not predict performance in other coursework or other desired professional competencies. For example, a listening comprehension test may not predict performance in some coursework (say humanities) to the same extent that it does in others (say, science, or vice versa), or in professional contexts such as listening skills in interpersonal situations, due to the way listening comprehension was assessed. This could be due, in part, to greater overlap in the way our construct—listening comprehension—was sampled in the test and course content in some domains, than in others. Because evidence on the technical adequacy of measurement are test and sample specific, guidelines for educational assessment call for obtaining technical adequacy evidence to support each new intended use and interpretation of assessment results (AERA, APA, & NCME, 1999).

Finally, our focus here has been on reliability. As noted previously, reliability and validity are not synonymous: Good evidence to support consistency in measurement—i.e., reliability—does not guarantee validity. For example, in education, it is not uncommon to hear reference to students' preferred modality of instruction (she's a visual learner; he's a kinesthetic learner, and so on). Although assessments may identify a preferred learning style the majority of the time, the instructional validity of this assessment information—i.e., whether assessment results inform instruction in ways that are associated with beneficial educational outcomes—is a separate issue (see Kavale & Forness, 1999). As education professionals, we are often—and should be—drawn to issues of validity, particularly, instructional and consequential validity. However, we should also appreciate the implications of reliability for professional decision making: Although good evidence of reliability does not equate to good evidence of validity, reliability likely places limits on the quality of our decision making (Macmann & Barnett, 1999). If assessment results are not consistent, the data that they provide may not be trustworthy and decisions based on that information may be misinformed.

How Does Classical Test Theory Differ From Modern Test Theory?

Modern test theory, and specifically item response theory (IRT), was developed independently along two different lines by different groups of scholars. Please refer to the chapters in this volume on IRT theory to see the issues involved here as well as the technical aspects of IRT. Here we are going to consider the essential, conceptual differences in these two approaches. For some of the testing that is done in higher education, CTT works quite adequately; for others, IRT is typically a preferable approach.

To appreciate the differences between the two approaches, we have to start by thinking about what they are trying to accomplish. Classical Test Theory looks at tests that are developed to measure traits in individuals, and tries to assess the dependability of results from those tests and help in improving these measures. It has a focus on the test and consistency of test results with a group of individuals.

Item Response Theory looks less at tests per se than it does at obtaining measures of individual performance and assessing the quality of that measurement. IRT starts with the notion that people's levels of performance are locatable along a continuum from low to high on a given construct, and that test items are locatable along that same continuum. If a person has the same proficiency as an item, that person has a 50% chance of getting that item right. In order to estimate the skill level of a person, we give that person some items, see how many they get right, and then see where they would have to be on the proficiency continuum in order for them to have had the best chance of getting the score that they obtained. This process also allows for the estimate of a standard error of measurement that is particular to this situation and this person's score. Please note that what is described here is pretty much the simplest version of IRT modeling. There are many wonderful elaborations on this model, but this model is probably used more often than not in real situations.

It requires a fair amount of work and a lot of testees to develop IRT scales in a given area. Once a set of items (item bank) exists that have been IRT "scaled," it is possible to administer basically any set of these items that seem appropriate to an individual and get an estimate of the person's skill level and an associated standard error. This has the distinct advantage of being able to tailor tests to the skill development of the individuals, and to the need for accuracy—if you need a smaller standard error in a given situation, you simply administer more items.

IRT, which is explained excellently elsewhere in this volume, has many wonderful qualities, and, when it is possible to execute, is almost always preferable to CTT. However, and this is a big "however" for higher education, it is not always possible, or even desirable, to do the work necessary to take an IRT approach to a given testing situation. A final exam in a course that does not enroll hundreds of students is such an example, as might a placement examination for courses in German where we simply want a rough idea of whether the student should take German 101 or German 102.

HOW DOES CLASSICAL TEST THEORY RELATE TO TESTING IN HIGHER EDUCATION TODAY?

For this final section, let's begin by returning to the examples discussed in the opening of this chapter.

Testing Before Admission

Let's look at college entrance examinations, tests used for selection and placement into courses, and tests used to assess achievement in courses. To begin, college entrance examinations have all gone toward IRT, at any rate in the United States. These are measures that require high levels of precision and consistency across different administrations of the measures, and IRT simply provides a much more robust statistical model for doing this than CTT.

Testing for Course Planning

The second type of testing in higher education that we considered was selection into courses, perhaps developmental or perhaps part of a regular sequence, once people have been accepted into a college or university. There are some commercially available measures for making such determinations, and they might use CTT or IRT. Other measures are often developed within the institution, and they are typically developed using CTT. This process usually involves a fair amount of time defining just what is meant by constructs to be measured. Do we want to include writing skills as part of a measure to determine placement into a remedial literacy course, or just reading skills? Do we want to use reading materials from course texts, or write new materials? How will we assess the students' skills once we have the reading materials? Will we let testees refer back to the reading material, or should we have them read it once and then respond based on their memory of the material. Once these issues have been decided, a rough draft of the test is generated that usually has more items than are ultimately needed, and it is "pilot tested" on a sample of individuals who are similar to those for whom the test is intended. The pilot results are examined to determine whether the scale seems to be internally consistent and tapping into one set of skills or more than one domain, item difficulty estimates and item to total score correlations. The items are culled and refined, and an iterative process, typically involving validity evidence in addition to reliability evidence, until a final form is produced. There are many good texts available on how to produce a good test; one we would recommend is Downing and Haladyna (2006).

Testing as Part of Course Assessment

Course examinations can involve a rather special use of CTT. In developing an examination for a course, we usually cannot give it a "pilot test" before actually using it. We can, however, use the data from a given examination to refine that exam for the next year's use, and to learn about the quality of our efforts in general. A number of statistical packages such as SPSS (see, for example, Norušis, 2008) or SAS (see, for example, Cody & Smith, 2005) are available to allow you to look at the overall reliability of a test, and how well each of the items related to the total score.

As mentioned above, there are issues with the notion of reliability and how high one wants it to be in any given situation. When looking at internal consistency reliability, what one is basically measuring is how strongly all of the items on a test intercorrelate with one another. If we think we are measuring a single trait or construct that is unidimensional in nature, then we might think that the higher the internal consistency reliability, the better. But if we are measuring achievement in History 101, then we might be agglomerating knowledge and skills that are somewhat differentiated one from another. We might be interested in some facts, the ability to critique an argument about an historical situation, and the ability to write a good short essay about a topic. There is no reason to believe that these should all necessarily be highly correlated with one another. Also, there might be items that are important in the course that almost all (or even all) of the students know and get correct. These items will not contribute at all to a higher reliability estimate, but it might be the case that we want to know that the students know them, and more importantly, we want to give students credit for knowing them (if they aren't on the test, we can't give credit for our students having learned them).

Thus, in this situation, which is probably the more common situation, we are not really looking for the highest possible coefficient alpha! In fact, that might indicate to us that we have too narrowly defined our examination. There is some work to indicate that we might want to conceptualize reliability in a somewhat different fashion here, one that looks more broadly at reliability as being a measure of the sufficiency of information that we have (Shepard, 2006; J. K. Smith, 2003).

Additional Considerations When Assessing Development and Learning in Higher Education

As noted previously, the historical roots of CTT evolved in the context of trying to measure unobservable, yet typically stable, traits, like extroversion or shyness (Christ & Hintze, 2007). When our assessments occur at one point in time, as in each of the above examples, such approaches to evaluating technical adequacy work fairly well. In education, however, we are often interested in learning and change over time. By way of example, suppose students were assessed at the beginning of the semester on a measure that sampled from material that they were expected to learn across the semester. We might make two predictions about our students' performance at this point: First, we might expect that as a group they might not do very well because they hadn't yet been taught the material on which they were being assessed. Second, we still might expect variability in performance between students, due to such factors as background knowledge and the like. If we were to reassess our students on this material across the term, we could make a third prediction, and that is, that we expect our students' performance to improve across the semester. Moreover, we could plot each student's performance over time and from this extrapolate their trajectory toward end of semester performance. Thus, it may be that some students didn't know course material at the beginning of the semester but they are now making good progress toward mastering material across the semester. Other students, however, may have done poorly on our assessment at the beginning of the semester and their rate of progress is slower. They are learning but their rate of progression is such that it is unlikely that they will successfully master course material if they don't pick up their pace of progress. As professionals, we may want to influence a student's learning trajectory, perhaps via an educational program or intervention, and evaluate not only whether the student's level of performance has changed but also whether we have enhanced their learning trajectory via participation in our programs or instructional interventions. In sum, we may want to know what to do to help our students and whether our efforts are effective.

These assessment purposes raise additional methodological considerations (see, for example, McCoach, Rambo, & Welsh [...]). Oftentimes, questions of effectiveness are equated with summative evaluation: Do program evaluations document effectiveness of the approach or practice? Considerations in designing effective program evaluations are beyond the scope of this chapter, and the interested reader is referred to Grayson.

Questions of what to do are, of course, questions of instructional utility. Assessment results and summative evaluation research on effective practice provide starting points but the uniqueness of learners and learning contexts limit the extent to which we can predict how a learner will respond to a given intervention or whether modifications would better serve student needs (Campbell & Overman, 1988). Two assessment approaches have been proposed to address this limitation: dynamic assessment (Caffrey, D. Fuchs, & L. S. Fuchs, 2008) and systematic formative evaluation (L. S. Fuchs & D. Fuchs, 1986).

In *dynamic assessment*, rather than presenting tasks to the student in one standardized condition, instruction or supports may be modified to observe student performance in different analogue learning conditions. Although dynamic assessment has been criticized as a fuzzy concept, scripted protocols may be developed and subsequently evaluated (D. Fuchs et al., 2007). Potential uses of dynamic assessment in higher education could include determining appropriate testing accommodations for students with disabilities (see Banerjee & Thurlow, this volume) and evaluation of strategies and supports that students might employ in the instructional context. When determining appropriate testing accommodations, the student might be presented with brief alternate forms of tests that incorporate the accommodations being considered, such as using a reader/writer or assistive technology to determine what is most appropriate (Braden & Joyce, 2008). In the instructional context, we might ask the student to perform an academic task, such as listening to a lecture excerpt or reading a text passage, under varying conditions and assess the student's performance for those conditions. Did the student demonstrate better listening comprehension when she sat back and listened to the lecture material? Or when she took narrative notes? Or when she took notes alongside a print-out of the lecturer's presentation slides? Did the student demonstrate better comprehension of textbook material when he followed along on an e-copy of the text as it was read aloud to him using text-to-voice software, or when he listened to an audio-recording of the text, or when he was taught pre-viewing and questioning strategies to use in approaching written texts?

Systematic formative evaluation, on the other hand, involves implementing the instructional modification in situ and assessing response following implementation of instructional intervention. Repeated assessments provide the opportunity to observe not only whether the student's level of performance has increased following intervention but also whether his learning trajectory has been enhanced (Christ & Hintze, 2007). Reflecting on progress monitoring results then informs decisions about whether instructional interventions are being effective in meeting student needs or whether modifications may be indicated to better meet learning needs (see Ervin & Schaughency, 2008; Schaughency, Alsop, & Dawson, 2010, for discussion).

Traditional approaches to measurement development that evaluate measurement by consideration of scores obtained at one point in time are still important when considering measurement tools for progress monitoring, but should be considered to be a starting point (L. S. Fuchs, 2004). In addition, we want to know whether our measures are sensitive to detecting and depicting growth and change (Hintze, 2009), whether our indices of growth and change are reliable (Christ & Hintze, 2007), and, if our results are to be used formatively, whether there is evidence to support instructional utility (L. S. Fuchs, 2004).

For reasons of efficiency, for such purposes, we may be interested in brief, repeatable indicators of relevant assessment targets. This measurement perspective is influenced by both behavior analytic and *curriculum-based measurement* approaches to assessment (Christ & Hintze, 2007; Deno, 2003). Most measurement development on this perspective has occurred in primary education settings. Work at the secondary level is underway (Twyman & Tindal, 2007), with scope for extension to higher education settings (Schaughency, Stanish, & Dickman, 2001; G. Tindal, personal communication, July 20, 2010).

Curriculum-based measurement falls under the general umbrella of curriculum-based assessment, defined as any set of measurement activities that record a student's performance in the local curriculum as a basis for gathering information to inform instructional decisions (Hintze, 2009). In higher education, this approach might be envisioned as assessing skills within the context of content area task demands (Behrman, 2000). Behrman and Street (2005), for example, describe using passages from a course text book to assess content-specific reading comprehension at the beginning of the semester to predict course grades.

In systematic formative evaluation, assessment probes would be administered in the context of ongoing instructional supports and accommodations to inform decisions about utility of those strategies and remaining learning needs. An example of systematic formative evaluation in higher education is provided by Schaughency et al. (2001), and illustrates the premises that (a) diagnostic assessment and evidence-based treatment outcome research may provide useful starting points for practice, but (b) progress monitoring may be warranted to refine accommodation or intervention plans.

Schaughency and colleagues worked with university students determined to be eligible for student disability services as students with Attention Deficit Hyperactivity Disorder (ADHD) and prescribed medication to assist with management of symptoms in the university context. They recognized that although medication may be considered to be an evidence-based treatment for ADHD, response to medication and appropriate medication dosage cannot be determined a priori, necessitating evaluation of response to medication to inform decisions about appropriate medication, dosage (American Academy of Pediatrics Subcommittee on Attention-Deficit/Hyperactivity Disorder and Committee on Quality Improvement, 2001), and remaining academic accommodation needs. Following the rationale that evaluation should encompass functional educational targets rather than focus solely on symptomology of the disorder (Pelham, Fabiano, & Massetti, 2005), measures included coding lecture notes for main points based on Spires (1993) and brief post-lecture quizzes adapted from Evans, Pelham, and Grudburg (1994–1995) (Schaughency & VanderVeen, 2000). Importantly, a first step in measurement development for this project entailed determining whether these brief indicators, collected across the semester, predicted end of semester course performance, with results suggesting that they did (Schaughency et al., 1997).

There are a number of potential benefits of systematic formative evaluation in higher education outlined by Schaughency et al. (2001). These include, but are not limited to, evaluation of the specific instructional support strategy. Engaging in the systematic formative evaluation process helps student and service providers better understand the relationship between the learner's characteristics, task demands of the university setting, and educational accommodations. Documentation of this process provides evidence of fulfillment of responsibilities by student and service providers, in the spirit of legislation relevant to providing services to students with disabilities (Bourke, Strehorn, & Silver, 2000; Foote, 2000; Shaw, McGuire, & Madaus, 1997).

Moreover, documentation may assist in accessing additional resources for students. For example, recognizing the association between higher education and employability (Graham & Stacey, 2002), state vocational rehabilitation agencies may support individuals on their caseloads who are pursuing higher education as was the case for one of the participants in the model demonstration project described by Schaughency and colleagues (2001). Upon receiving documentation that the participant experienced a specific learning disability and benefited from use of assistive technology, the state vocational rehabilitation agency supplied a laptop and specialized software to assist with the student's university coursework. In addition, the agency contacted the principal investigator to ask if she would provide these evaluation services for other clients on their caseload, commenting on the perceived helpfulness of the information provided to their decision making about allocation of resources. Post-hoc support for the agency's investment in resources for the participating student is provided by informal follow up with this student who had previous false starts at undergraduate education and a transcript peppered with grades ranging from As, hinting at his competencies, to less successful performance. Not only did he successfully complete his undergraduate degree following participation, but at last contact was enrolled in PhD study in the area of assistive technology in special education.

There is also the possibility of positive reactive effects to systematic formative evaluation. For the student, the process of participation and ongoing objective feedback can facilitate development of self-regulated learning and self-determination, important developmental tasks in higher education for all students including those with disabilities (Wehmeyer & Schwartz, 1997). Moreover, engaging in systematic formative evaluation may help faculty become aware of issues and strategies, which are then incorporated into teaching practices, thereby yielding positive reactive effects that emanate beyond the target student to potentially benefitting all students. For example, developing classroom-based assessment probes necessitated that faculty identify the main points to be covered in lectures. In so doing, one faculty member commented that it would be probably good to explicitly articulate this information to students in class, and subsequently began to do so.

To conclude, systematic formative evaluation and dynamic assessment are assessment practices designed to enhance instructional utility by examining response to intervention, and ultimately enhance outcomes for students (consequential validity). Assessing the moving target of learning and development over time or in response to intervention, however, raises additional measurement considerations. Given the importance of decisions based on these assessment results, evidence is needed to support the dependability of the assessment results used to inform these decisions.

SUMMARY

Classical test theory has been a stalwart in higher education over the decades. Its influence has waned dramatically and appropriately in many areas of higher education, but is still useful for specific applications in higher education, and very much so for measures that are associated with specific courses. Understanding CTT also provides a solid introduction to understanding the technical issues related to testing in general. We hope this chapter has facilitated such an understanding.

NOTE

1 This chapter represents a collaboration of members of the Literacy Research Unit at the University of Otago. The Literacy Research Unit, housed in the Educational Assessment Research Unit of the College of Education, was formed as a partnership between faculty in the College of Education, Higher Education Development Centre, and Department of Psychology to foster a research community dedicated to learning about the literacy needs of students in higher education and to inform evidence-based practice for addressing those needs in tertiary settings. This chapter was supported in part by a grant from the University of Otago to the Literacy Research Unit. Material in the chapter referring to systematic formative evaluation in the university setting was supported by a grant from the U.S. Department of Education, Office of Special Education Programs (CFDA Award H078C6005). Opinions expressed herein are those of the authors and do not necessarily

represent the opinion of the U.S. Department of Education. We would also like to acknowledge the students, faculty, and other higher education service providers with whom we have worked over the years for shaping our thinking about the importance of good data for educational decision making in higher education.

REFERENCES

American Academy of Pediatrics Subcommittee on Attention-Deficit/Hyperactivity Disorder and Committee on Quality Improvement. (2001). Clinical practice guideline: Treatment of the school-aged child with attention-deficit/hyperactivity disorder. *Pediatrics, 108*, 1033–1044.

American Educational Research Association, American Psychological Association, & National Council on Measurement in Education. (1999). *Standards for educational and psychological testing*. Washington, DC: American Educational Research Association.

Banerjee, M., & Thurlow, M. L. (this volume). *Using data to find common ground between secondary and postsecondary accommodations for students with disabilities.*

Behrman, E. H. (2000). Developmental placement decisions: Content-specific reading assessment. *Journal of Developmental Education, 23*, 12–17.

Behrman, E. H., & Street, C. (2005). The validity of using a content-specific reading comprehension test for college placement. *Journal of College Reading and Learning, 25*, 5–22.

Bourke, A. B., Strehorn, K. C., & Silver, P. (2000). Faculty members' provision of instructional services to students with LD. *Journal of Learning Disabilities, 33*, 26–32.

Braden, J. P., & Joyce, L. B. (2008). Best practices in making assessment accommodations. In A. Thomas & J. Grimes (Eds.), *Best practices in school psychology V* (Vol. 2, pp. 589–604). Bethesda, MD: National Association of School Psychologists.

Brennan, R. L., & Kane, M. T. (1979). Gerneralizability theory: A review. In R. E. Traub (Ed.), *Methodological developments* (New Directions for Testing and Measurement, No. 4, pp. 33–51). San Francisco, CA: Jossey-Bass.

Caffrey, E., Fuchs, D., & Fuchs, L. S. (2008). The predictive validity of dynamic assessment: A review. *The Journal of Special Education, 41*, 254–270.

Campbell, D., & Overman, E. S. (Eds.). (1988). *Methodology and epistemology for social science: Selected papers by Donald T. Campbell*. Chicago. IL: University of Chicago Press.

Christ, T. J., & Hintze, J. M. (2007). Psychometric considerations when evaluating response to intervention. In S. R. Jimerson, M. K. Burns, & A. M. VanDerHeyden (Eds.), *Handbook of response to intervention: The science and practice of assessment and intervention* (pp. 93–105). New York, NY: Springer.

Cody, R., & Smith, J. K. (2005). *Applied statistics and the SAS programming language* (5th ed.). Englewood Cliffs, NJ: Prentice-Hall.

Cronbach, L. J. (1951). Coefficient alpha and the internal structure of tests. *Psychometrika, 16*, 297–334.

Deno, S. L. (2003). Developments in curriculum-based measurement. *The Journal of Special Education, 37*, 184–192.

Donlon, T. F. (Ed.). (1984). *The College Board technical handbook for the Scholastic Aptitude Test and Achievement Tests*. New York, NY: College Entrance Examination Board.

Downing, S. M., & Haladyna, T. M. (2006). *Handbook of test development*. Mahwah, NJ: Erlbaum.

Drezek McConnell, K., & Doolittle, P. E. (this volume). *Classroom-level assessment: Aligning pedagogical practices to enhance student learning.*

Ervin, R. A., & Schaughency, E. (2008). Best practices in accessing the systems change literature. In A. Thomas & J. Grimes (Eds.), *Best practices in school psychology V* (Vol. 3, pp. 853–874). Bethesda, MD: National Association of School Psychologists.

Evans, S. W., Pelham, W., & Grudburg, M. V. (1994–1995). The efficacy of notetaking to improve behavior and comprehension of adolescents with attention deficit hyperactivity disorder. *Exceptionality, 5*, 1–17.

Feldt, L. S., & Brennan, R. L. (1989). Reliability. In R. L. Linn (Ed.), *Educational measurement* (3rd ed., pp. 105–146). New York, NY: American Council on Education.

Foote, W. E. (2000). A model for psychological consultation in cases involving the American with Disabilities Act. *Professional Psychology: Research and Practice, 31*, 190–196.

Fuchs, D., Fuchs, L. S., Compton, D. L., Bouton, B., Caffrey, E., & Hill, L. (2007). Dynamic assessment as responsiveness-to-intervention: A scripted protocol to identify young at-risk readers. *Teaching Exceptional Children, 39*, 58–63.

Fuchs, L. S. (2004). The past, present, and future of curriculum-based measurement research. *School Psychology Review, 33*, 188–192.

Fuchs, L. S., & Fuchs, D. (1986). Effects of systematic formative evaluation: A meta-analysis. *Exceptional Children, 53*, 199–208.

Geisinger, K., Shaw, L. H., & McCormick, C. (this volume). *The validation of tests in higher education.*

Graham, P. A., & Stacey, N. G. (2002). *The knowledge economy and postsecondary education: Report of a workshop.* Washington, DC: National Academy Press.

Grayson, T. E. (this volume). *Program evaluation in higher education.*

Guttman, L. A. (1945). A basis for analyzing test-retest reliability. *Psychometrika, 10*, 195–200.

Hintze, J. (2009). Curriculum-based assessment. In T. B. Gutkin, & C. R. Reynolds (Eds.), *The handbook of school psychology* (4th ed., pp. 397–409). Hoboken, NJ: Wiley.

Hoyt, C. (1941). Test reliability obtained by analysis of variance. *Psychometrika, 6*, 153–160.

Hunsley, J., & Meyer, G. J. (2003). The incremental validity of psychological testing and assessment: Conceptual, methodological, and statistical issues. *Psychological Assessment, 15*, 446–455.

Kavale, K. A., & Forness, S. R. (1999). Effectiveness of special education. In C. R. Reynolds, & T. B. Gutkin (Eds.), *The handbook of school psychology* (3rd ed., pp. 984–1024). New York, NY: Wiley.

Kuder, G. F., & Richardson, M. W. (1937). The theory of the estimation of test reliability. *Psychometrika, 2*, 151–160.

Linn, R. L. (1994). Performance assessment: Policy promises and technical measurement standards. *Educational Researcher, 23*, 4–14.

Lord, F. M., & Novick, M. R. (1968). *Statistical theories of mental test scores.* Reading, MA: Addison-Wesley.

Macmann, G. M., & Barnett, D. W. (1999). Diagnostic decision making in school psychology: Understanding and coping with uncertainty. In C. R. Reynolds & T. B. Gutkin (Eds.), *The handbook of school psychology* (3rd ed., pp. 519–548). New York, NY: Wiley.

McCoach, D. B., Rambo, K., & Welsh, M. (this volume). *Issues in the analysis of change.*

Morgan, D. L. (this volume). *College placement testing of entering students.*

Norušis, M. J. (2008). *SPSS 16.0 guide to data analysis.* Upper Saddle River, NJ: Prentice Hall.

Pelham, W. E., Jr., Fabiano, G. A., & Massetti, G. M. (2005). Evidence-based assessment of attention deficit disorder in children and adolescents. *Journal of Clinical Child and Adolescent Psychology, 34,* 449–476.

Pitoniak, M. J., & Morgan, D. L. (this volume). *Setting and validating cut scores for tests.*

Schaughency, E., Alsop, B., & Dawson, A. (2010). The school psychologist's role in assisting school staff in establishing systems to manage, understand, and use data. In G. G. Peacock, R. A. Ervin, E. J. Daly, III, & K. W. Merrell (Eds.), *Practical handbook of school psychology: Effective practices for the 21st century* (pp. 548–565). New York, NY: Guilford Press.

Schaughency, E., Stanish, H., & Dickman, H. (2001). **A**dvancement through **C**ooperative **C**ollege **E**ducation for **S**tudent **S**uccess (ACCESS): Evaluating accommodation utility and empowering the student with disabilities. Allendale, MI: Grand Valley State University. Available upon request from schaughe@psy.otago.ac.nz

Schaughency, E., Stanish, H., Dickman, K., Krum, K., Pedraza, J., & Rodriguez, S. (1997, June). *Developing ecologically valid measures to assess postsecondary students' performance in the classroom.* Paper presented at the National Transition Alliance for Youth with Disabilities 12th Annual Project Directors Meeting, Washington, DC.

Schaughency, E., & VanderVeen, K. (2000, July). *Selecting accommodations, designing environments, and empowering students through data-based decision-making (DBDM).* Paper presented at the annual meeting of the Association for Higher Education and Disability, Kansas City, MO.

Shaw, S. F., McGuire, J. M., & Madaus, J. W. (1997). Standards of professional practice. *Journal of Postsecondary Education and Disability, 12,* 26–35.

Shepard, L. A. (2006). Classroom assessment. In R. L. Brennan (Ed.), *Educational measurement* (4th ed., pp. 623–646). Westport, CT: American Council on Education/Praeger.

Smith, G. T. (2005). On construct validity: Issues of method and measurement. *Psychological Assessment, 17,* 396–408.

Smith, J. K. (2003). Reconceptualizing reliability in classroom assessment. *Educational Measurement: Issues and Practice, 22*(4), 82–88.

Spires, H. (1993). Learning from a lecture: Effects of comprehension monitoring. *Reading Research and Instruction, 32,* 19–30.

Thompson, B., & Vache-Haase, T. (this volume). *Reliability.*

Twyman, T., & Tindal, G. (2007). Extending curriculum-based measurement to middle/secondary schools. *Journal of Applied School Psychology, 24,* 49–67.

Webb, N. M., Shavelson, R. J., & Steedle, J. T. (this volume). *Generalizability theory in assessment contexts.*

Wehmeyer, M., & Schwartz, M. (1997). Self-determination and positive adult outcomes: A follow-up study of youth with mental retardation or learning disabilities. *Exceptional Children, 63,* 245–255.

Zwick, R. (this volume). *Admissions testing in higher education.*

Reading 3.2

STANDARD ERROR OF MEASUREMENT (SEM)

Melissa Fein

The standard error of measurement (SEM) is used to create a score band around an individual's score. This score band takes into account that one exam score is just one estimate that is subject to error, and that the examinee's true score is better captured by a range of values expressed within a probability statement. This section of the technical appendix provides instructions for computing the SEM, includes directions for construction of an error band, and discusses the implications of the SEM for CRT development.

COMPUTATION

To compute the standard error of measurement the following steps need to be followed:

- Compute an appropriate reliability coefficient.
- Compute the test score standard deviation.
- Plug the reliability coefficient and the test score standard deviation into the SEM formula given below:

$$\text{SEM} = sd\sqrt{1-r}$$

Where: sd = standard deviation and r = reliability.

Computation of reliability and standard deviation are presented in other sections of the appendix and are not repeated here.

Example 1:

- The standard deviation of the test scores is computed to be 11.8.
- The reliability is computed to be .82.

$$\text{SEM} = sd\sqrt{1-r} = 11.8\sqrt{1-.82} = 5.01$$

To compute this using a spreadsheet, follow these steps:

- Enter and label the standard deviation and reliability values as given in row 1 and row 2 of Table 3.1.
- Use cell A3 to label the result. Type SEM in cell A3.

- Place the cursor in cell B3. This is where the computed result will be displayed.
- In cell B3, type: = B1*SQRT(1–B2) and press enter.

As in the hand-computed example, the resulting standard error is 5.01.

TABLE 3.1

	A	B
1	Standard deviation	11.8
2	Reliability	.82
3	SEM	5.0063
4		

ERROR BANDS

A rough way to use the standard error of measurement is to say that the score the examinee earns on the exam is just a point estimate subject to error, and that the examinee's true score falls, approximately, within a range of 2 SEM [(2)(5.01)] is approximately 10 points above and below the mean. So if an examinee earns a score of 75 on an exam, his true score range is between scores of 65 and 85 points. A more formal way of expressing this idea is to create a precise confidence interval or confidence band.

To compute a formal confidence band, the standard error is not multiplied by 2 as in the informal example, but is multiplied by:

- 1.96, for a 95 percent confidence band; or
- 1.65, for a 90 percent confidence band.

In this example, the 95 percent score band for a student with a 70 is:

- Lower limit: 70 – (1.96)(5.01) = 60.18
- Upper limit: 70 + (1.96)(5.01) = 79.82

So taking measurement error into account, the probability is 95 percent that this student's true competency score falls within the score range of 60.18 to 79.82.

IMPLICATIONS

There are implications of banding with respect to the cut score. If the cut score is 60, even the lower bound of the examinee who scores a 70 falls above this cut, so his score band overlaps the passing range. But what about the examinee who scores just barely below the cut point? The upper limit of his score band will fall within the passing zone. Should he pass or fail? A comprehensive coverage of banding issues is beyond the scope of this book. But this highlights the importance of trying to build a high-quality instrument with reliable scores.

The following illustrates the effect that a change in reliability has on the SEM, when everything else remains the same.

When reliability is 0.60: $\text{SEM} = sd\sqrt{1-r} = 11.8\sqrt{1-0.60} = 7.46$

When reliability is 0.80: $\text{SEM} = sd\sqrt{1-r} = 11.8\sqrt{1-0.80} = 5.28$

When reliability is 0.95: $\text{SEM} = sd\sqrt{1-r} = 11.8\sqrt{1-0.95} = 2.6$

When reliability is better, or higher, the SEM is smaller, which means the score band will be tighter, or narrower.

It should be noted that when the number of examinees is small, the standard deviation is not going to be stable. This means it will probably be quite different for another sample of examinees. It is a best practice guideline to track the SEM. However, when small-scale testing programs are first getting started and the number of examinees is small, the SEM is not always helpful and care must be taken in making interpretations.

The directions above are for computing a general SEM. Computation of the SEM at a cut score is outside the scope of this book.

HOW DOES CLASSICAL TEST THEORY APPLY TO MY PROFESSION?

As a classroom teacher or educational leader, it is important to realize that all assessments have error and the results are affected by error. The challenge is to reduce error by using effective measurement techniques as well as other things (statistical analyses, test blueprints, content experts, appropriate environments, etc.). There are both random and systematic sources of error. Random sources can be mood, health, level of anxiety, motivation, and so on. As an educator, many of these sources are out of your control; however, you can use practices that promote healthy lifestyles, reduce anxiety, increase motivation, and increase positive moods. Much of the systematic error is in the control of the teacher. These factors often include directions on the assessment, the actual wording of the items, temperature of the room, and noise in and around the room. A fundamental yet extremely important concept in CTT is the fact that all assessments include error; therefore, there is no way (in the real world) to perfectly measure what somebody knows or can do. The score that a student receives on an assessment is simply a numeric value that humans created to quantify knowledge, skills, abilities, and so on. Keep in mind that you are making decisions about students, teachers, or even the school that have significant impacts on those affected by those decisions. So, please remember the effects of measurement error when you are making consequential decisions.

CHAPTER 4

Types of Assessments

Formative, summative, standardized, teacher- and leader-created, achievement, personality, aptitude, intelligence, and interests: Why so many?

AS YOU READ

As you read the chapter, think about the following:

- How do **you** define or explain various types of assessments?
- What is **your** experience with using various types of assessments?

WHAT ARE THE GENERAL PRINCIPLES OF VARIOUS TYPES OF ASSESSMENTS?

Formative Versus Summative Assessment

To be honest, there is no one single definition of each of these types of assessments. Instead, they are generally labeled as one or the other based on the characteristics of the assessment and more importantly the purpose of the assessment. Here are some general guidelines that help categorize each type of assessment. Think about the following questions:

- What is the purpose of the assessment? If the answer to this question is to determine a grade, then the assessment is likely summative. If the purpose is to drive instruction, then the assessment is likely formative.
- What will I do with the results? If the answer to this question is to assign a grade, then the assessment is likely summative. If the purpose is to determine what to teach or reteach, then the assessment is likely formative.

As you can see, formative assessments drive curriculum and instruction and summative assessments help you determine what grade to give a student or to make a judgment about learning. In both cases, formative and summative assessments provide information to determine what students know and what they can do. When using formative assessments, you use the information you obtain to determine what you need to teach and perhaps even what instructional strategies to use. Formative assessments occur throughout the lesson or unit. These results inform the teacher how to move forward and at what pace. The results of formative assessments provide opportunities for students to receive feedback about their knowledge and/or skills with the absence of consequences. The process also informs the classroom teacher about what concepts they need to revisit to increase understanding.

Conversely, when using results of the assessment in a summative nature, you typically assess students' knowledge and/or skills at the end of a lesson or unit. These results then help inform final decisions about students' knowledge and/or skills based on curriculum standards.

Other than the timing of the assessment and the purpose of the assessment, the characteristics of the assessments may also differ. Often, formative assessments are brief and there are no consequences associated. Students receive feedback about their knowledge and/or skills. Student performances on summative assessments often have associated consequences (grades, grade retention/promotion, etc.).

The primary difference between formative and summative assessment is how you use the results. The instruments themselves may be the same. But, the use of the results is very different between the two. Both types of assessments are equally important to the learning process. Formative assessment allows the teacher and students to learn about what they know and do not know as a means for guiding additional instruction. Summative assessment informs educators of the status of students' knowledge and/or skills at the end of the predetermined instructional time (lesson and/or unit) (Harlen, n.d.; Hill et al., 2018; Ridhwan, 2017; Taras, 2005).

Standardized Tests

Standardized tests are extremely prevalent in our accountability-based societal systems—especially in education. Standardized tests are large-scale assessments that have undergone extensive development, analyses, norming and/or scaling of scores, and so on. Typically, there are multiple types of standardized tests that which you may already be familiar. The following are the most common types of standardized tests:

- **Achievement tests**: Measure what an individual knows
- **Personality tests**: Measure traits and characteristics of an individual (e.g., attitudes, values)
- **Aptitude tests**: Assess an individual's potential or future performance (e.g., predict future job performance)

- **Intelligence tests**: Measure a latent construct; typically a group of related variables (e.g., verbal, comprehension, memory) that we call "intelligence" or "IQ," which refers to an intelligence quotient
- **Interest inventories**: Typically used to measure at what an individual is best or for what he is best suited (e.g., career choice)

Each of you is likely familiar with all or at least most of these types of standardized tests. In our education system, achievement tests are extremely popular and are used to measure student knowledge, teacher's performance, and school rankings. Using standardized tests to make high-stakes decisions for students (e.g., grade retention), teachers' evaluations (e.g., pay for performance), and school rankings (e.g., funding opportunities and/or cuts) is extremely controversial. There is a whole body of literature on this topic and it is one that has personally interested me for quite some time. In fact, the literature review in my dissertation emphasized the history of the accountability movement as well as some of the controversy that surrounds the topic.

Achievement Tests

Achievement tests measure what a student knows. In education, achievement tests are the most common form of standardized testing. Achievement tests fall into one of two types of tests, criterion-referenced and norm-referenced.

Criterion-Referenced Tests and Scoring

In an accountability era, stakeholders often ask the question, "What do students know?" Many states use the results of criterion-referenced tests to answer this question. **Criterion-referenced tests** are intended to measure students' knowledge as it related to specific criteria. Curriculum standards are typically the framework for determining how a student performs on specific curriculum standards. For example, Common Core Standards Initiative was an attempt to create nationwide standards outlining what each child should know at specific grade levels throughout his schooling. These standards provide frameworks for educators across the nation. At designated times, students are assessed on their knowledge of each of the standards. This type of assessment is considered criterion-referenced because the students' performance on the assessment is aligned to the standards (or criteria). Scores are typically reported in percentage correct and/or scales. We will discuss scale scores in depth in a later chapter.

Norm-Referenced Tests and Scoring

Norm-referenced tests differ from criterion-referenced tests in that students' performance on a test is not an indication of his knowledge of standards. Instead, their performance is compared with a group of like students that completed the same assessment. For example, a group of fourth-grade students in one setting takes the same exam as fourth-grade students in another setting. The data are analyzed in group form and various statistics are used to compute information about performance of the entire group of students on the test. Then, each student is compared with the entire group of students. Comparing the individual student performance with the entire group of students is considered norm-referencing: How Child A compares with the group of students that completed the same exam. This is called norm-referencing because the group of students is considered the normed group. Scores are typically presented in percentiles. Thus, when you review test scores that indicate, "Student A scored in the 70th percentile," you are typically dealing with normed-reference tests. Essentially, this statement means that Student A scored as well or better than 70 percent of all students in the normed-group that took the same exam. We will discuss percentiles in depth in a later chapter.

Personality Tests

Personality tests are less common in education than in other environments. Although they are less common, it is still important for you to understand what they are and why they are administered. **Personality tests** measure traits and characteristics of individuals. Often, employers use the results of personality tests to help them gauge whether an individual is a good fit for a particular job. Also, these tests are used with middle and/or high school students for students to better understand themselves as they try to narrow their perspective academic and/or employment options. Although there are multiple personality tests, one of the most common personality tests is the Myers-Briggs. In this case, the tests are neither criterion-referenced nor norm-referenced. Instead, the results indicate individual characteristics and traits.

Aptitude Tests

Aptitude tests are commonly administered in secondary and postsecondary education settings. These tests are designed to measure individual capabilities and abilities in predetermined constructs such as analytical reasoning, mechanical knowledge, verbal comprehension, and numerical operations. Often, these tests are used to determine how people react to situations and perform various skills. There are different types of aptitude tests including job skills aptitude and college-readiness aptitude. One of the most commonly used aptitude tests in education is the Scholastic Aptitude Test 1 (SAT 1). The SAT 1 measures students' aptitude in their verbal, mathematics, and writing. SAT 2 is different as it is an achievement test.

Intelligence Tests

Intelligence tests measure a latent construct; typically, a group of related variables (e.g., verbal, comprehension, memory) that we call "intelligence" or "IQ" which refers to an intelligence quotient. Generally, the results of intelligence tests are compiled from multiple sub-standardized tests that measure various constructs such as linguistic abilities, logic, verbal reasoning, special reasoning, quantitative reasoning, memory, and judgment. There are thought to be multiple intelligences so often the test that is selected to measure those intelligences is determined by the test administered. The culmination of performance on each of the sub-tests is calculated in one intelligence quotient (IQ). Three commonly administered IQ tests are the Stanford-Binet, Woodcock-Johnson test of cognitive abilities, and the Wechsler Intelligence Scale. In 1916, Lewis Terman designed the scale for quantifying intelligence including the following: Above 140, genius or near genius; 120–140, very superior intelligence; 110–119, superior intelligence; 90–109, normal or average intelligence; 8–89, dullness; and 70–79, borderline deficiency.

Interest Inventories

Interest inventories are typically used to measure at what individuals are best or for what they are best suited. Interest inventories are less common in educational settings than in other settings; however, similar to personality tests, interest inventories are occasionally administered with middle and high school students to determine possible career trajectories based on their interests or preferences. There are multiple interest inventories available for public use and it is common for educators to develop their own based on the purpose of giving the interest inventory. Just remember to follow instrument development guidelines for ensuring quality when developing your own.

Teacher- and Leader-Created Assessments

Often, as educators, we create our own assessments because who knows our students better than us? Creating your own assessments is certainly not a novel or new concept in education. However, creating your own assessments is one of the most difficult things that you can do in education. There can also be tremendous consequences if you do not create a good test. Teacher-created assessments are those that you create the items themselves. There are very specific guidelines to writing items and there are various ways to determine if the items actually assess what they are intended to assess. In a later chapter, we will discuss in detail how to write quality test items as well as how to determine the level of quality for each item and the overall test.

WHAT DOES THE LITERATURE SAY ABOUT VARIOUS TYPES OF ASSESSMENTS?

Reading 4.1

FRAMES OF REFERENCE

Interpreting Test Scores

Malcolm L. van Blerkom

INTRODUCTION

[Previously we] differentiated between measurement and evaluation. When you score a test, a quiz, or a similar device you are using measurement. You come up with a number or score that represents how well the student performed on that particular instrument. The next step is to find some way to interpret that score. Did the student perform well, about average, or poorly on that instrument? Was the student's score acceptable? This chapter will describe several frames of reference that you can use to interpret test scores. We will be discussing four frames of reference: ability-referenced, growth-referenced, norm-referenced, and criterion-referenced interpretations. We will discuss the advantages and disadvantages of each approach.

FOUR FRAMES OF REFERENCE

Whenever you give your students a test, quiz, or any other form of assessment instrument, you need to score that instrument. Many of our scoring systems result in a number. Let's say that one of your students, Charisa, received 76 points on her last math test. Of course, Charisa may want to know what that 76 stands for. Does it mean that she did well, about average, or poorly? There are a number of ways of looking at her test performance. Did she do about as well as we could expect based on her ability with mathematics? Does this test score reflect a growth in her skills since she took her last math test? How did she do in comparison to the other members of her 8th-grade math class? Does the test tell you specifically what new skills Charisa has mastered? Each of these questions reflects a different frame of reference, a different way to interpret her test score.

Ability-Referenced Interpretations

One approach used to interpret a test score is known as ability-referenced interpretations. With *ability-referenced* interpretations the student's test performance is compared to what we believe the student should be able to do based on his or her ability. We expect a high-ability student to perform better on a test than would an average-ability student. Although this approach sounds quite reasonable, it is highly problematic.

Definition
With **ability-referenced** interpretations the student's test performance is compared to what we believe the student should be able to do based on his or her ability.

The primary difficulty with this approach is that it is very difficult to obtain a good estimate of a student's ability. Many of us grew up believing that I.Q. tests and other ability tests are able to somehow magically tap into our innate abilities and measure them accurately. It is likely that in the early years when we first started using ability tests, even psychologists felt that those tests could very accurately tap into and measure our abilities. However, now that we have had over 100 years of experience with ability testing, we have come to recognize that scores on ability tests need to be interpreted cautiously. There are many variables that affect our performance on an ability test besides our abilities. For example, such tests generally assume that a student has had the same developmental experiences as other students. Your performance on a math ability test assumes that you have had typical educational experiences with mathematics. We know, however, that not all elementary teachers emphasize mathematics instruction equally. Especially in the very early grades, some elementary teachers spend less than 30 minutes a day teaching arithmetic whereas they spend nearly 3 hours teaching language arts skills (Stevenson, Stigler, & Lee, 1986). If you are dealing with a 3rd-grade student who has had relatively little arithmetic instruction in the past, that student's score on a mathematics ability test will surely underestimate that student's ability with math. Other variables that tend to depress ability test scores include poverty, disabilities, and language. Students who grow up in poverty or who have even a minor disability (or an illness) frequently score too low on ability tests. Students who grew up speaking another language, and only later learned English, tend to do less well than they should on a test given in English. For these students, their test scores frequently do not accurately reflect their abilities. The advice we frequently give today is to interpret ability tests cautiously. With many students, ability tests will underestimate a student's ability.

There is an additional problem with the ability-referenced approach. Many skills we teach in school are dependent on several different abilities. There is no single ability involved in performing well in social studies. We do not really know all of the skills involved with effective learning in social studies and many other areas. Therefore, we have no real way to estimate how a student should perform in most subject-matter areas.

Because of these and other problems we do not typically recommend that teachers use ability-referenced test interpretations.

Growth-Referenced Interpretations

A second approach to interpreting test scores is a comparison approach. With *growth-referenced* interpretations we compare the student's test score after instruction with the score from a similar test given prior to instruction. Essentially, it is a before-and-after approach, sometimes known as a pre-test/post-test approach. The first test is given prior to instruction and demonstrates what the student is able to do before being taught the new material. The second test is given after the instruction. A comparison of the two tests should, therefore, provide a clear picture of what the student has learned. Sometimes we even use the same test on each occasion and then compute a difference score. We assume that a student who scored 37 points more on the second test than the first test learned more than the student whose score only improved by 22 points. This appears very logical and sounds like a reasonable way to interpret test scores.

Definition

With **growth-referenced** interpretations we compare the student's test score after instruction with the score of a similar test given prior to instruction.

In reality, however, there are several serious problems with this approach. The first problem arises because difference scores are notoriously unreliable. All test scores contain what we refer to as measurement error [...]. Essentially, measurement error involves all those factors that cause a student to perform on a test in a way other than that in which he or she should have performed. All tests contain some measurement error. When we compute a difference score, as we frequently do with growth-referenced interpretations, we are computing the difference between two test scores. Unfortunately, the difference scores inherit the measurement error from each test and contain even more error than either original test did individually. This additive property of the measurement error is even more severe if the two tests are positively correlated with one another, as they often are since they are each measuring the same skill. As a result, difference scores frequently have so much measurement error that we can have little confidence in what they are telling us.

The second problem with growth-reference interpretations has to do with the nature of the learning curve. The learning curve tends to be non-linear; it is

S-shaped. If the learning curve were linear—a straight line—each day, as we were taught a new skill, our performance would improve by the same amount from one day to the next. However, that almost never happens. With just about any type of learning that we have studied over the years we get something quite different. When we start learning a new skill the daily increments in our performance are quite small. It frequently takes us a while to really catch on. After a while we see the learning curve accelerate and we seem to make tremendous strides in our performance from day to day. Finally, the learning curve begins to flatten out and we reach what has sometimes been referred to as a plateau. We seem to improve very little from day to day.

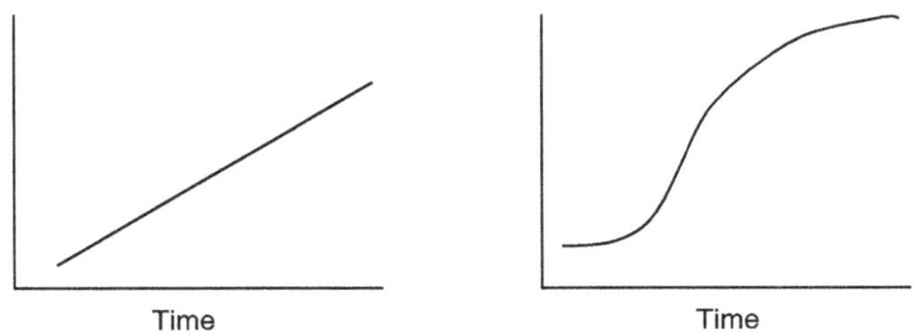

FIGURE 4.1 **A comparison of a linear learning curve with an actual S-shaped learning curve.**

If we try to compare two students using a growth-referenced interpretation, the learning curve can make interpretations problematic. Let's say that we have two 10th grade science students. George showed a 37-point improvement from his pre-test to his post-test. His friend, Rick, only displayed a 22-point improvement. We might be inclined to suggest from these test scores that George has learned more than Rick. However, if George and Rick started out at different points on the learning curve, such an interpretation is problematic. Let's say that George was at the acceleration point in the learning curve. We should expect to see substantial growth in his performance. On the other hand, if Rick were approaching the plateau part of the learning curve we should expect much less growth. Each student may have grown exactly as much as we should have expected him to grow. In reality, we never really know where a student is on the learning curve.

Since these two problems interfere with our ability to make good, sound interpretations of growth scores, we tend to use them very little.

Norm-Referenced Interpretations

A third way to interpret test scores is to compare a student's score on a test to the scores of other students who took the same test. With *norm-referenced* interpretations we compare the score a student received on a test with the scores from some norm group. In the case of a nationally standardized test, that norm group typically consists of a random selection of hundreds of other similar students who took the same test. For a classroom test, that norm group typically is the other students in the class. With norm-referenced interpretations, essentially the students who took the test are ranked from those with the highest scores to those with the lowest scores. The student's letter grade (or other evaluative index) is based on where he or she is on the list. If a student is near the top of the list, he or she gets an "A" or some other positive evaluation. If the student is near the bottom of the list, he or she receives a lower grade or some other lower evaluation (e.g. unsatisfactory).

Definition
With **norm-referenced** interpretations we compare the score that a student received on a test with the scores from some norm group.

Educators in the United States have used norm-referenced interpretations for several centuries. Until the 1970s, the vast majority of classroom tests and other instruments were evaluated with a norm-referenced interpretation. Although criterion-referenced interpretations have gained popularity and are used much more often today than they were in the past (for reasons that we will discuss later), norm-referenced interpretations are still considered the most appropriate approach in many situations.

Unfortunately, teachers who use norm-referenced interpretations frequently fail to take an important step. When making a norm-referenced interpretation, it is extremely important to fully describe the norm group. Since it is based on a comparison, it is essential that you describe the group with whom the student is being compared. For example, if Tanya received an "A" for language arts that was based on a norm-referenced interpretation, it could mean that Tanya is one of the very best students in language arts in a class where most of the students are quite competent. However, it could also mean that Tanya was one of the best students in a class where the vast majority of the students actually displayed low levels of competence. Descriptions of the norm group would be especially important if the norm group is small. Larger norm groups are more likely to mimic the population. Unfortunately, norm-reference interpretations frequently do not tell us very much about what skills the student has mastered.

Criterion-Referenced Interpretations

The fourth way to interpret test scores is to compare each student's score to some preset standard or criterion. This is known as *criterion-referenced* interpretation. This could be a set of criteria such as those we used when I was in high school. To earn an "A" you needed to score between 93% and 100%; any score from 85% to 92% resulted in a "B"; scores between 78% and 84% resulted in a "C", and scores from 70% to 77% resulted in a "D." The standard could also be set up on a pass/fail basis. On the college campus where I work, our algebra review course is set up so that a final average of 75% is required to pass the course. Any student who scores below 75% fails the course. The Praxis test is used as a teacher certification test in many states. In Pennsylvania, college students must earn a score of 173 on the mathematics test to pass.[1] There are many different ways in which we can set up criteria.

Definition

With **criterion-referenced** interpretations a student's test score is compared to some pre-set standard or criterion.

Criterion-referenced interpretations started to become more popular in the United States in the 1970s, largely in response to the criticisms leveled at norm-referenced interpretations. The major advantage of criterion-referenced interpretations is that they provide a clearer view about which skills the student has mastered. Since many educators found that this was highly desired, they were persuaded to use criterion-referenced interpretations more frequently. We will compare norm-referenced and criterion-referenced interpretations later in this chapter.

Many teachers believe that they are using criterion-referenced interpretations, but fail to meet an essential requirement to do so. When using such interpretations, it is vital to spell out clearly the domain that the test is covering. For example, if you were to give your 6th-grade class a test based on the last chapter that you covered on the economy of the Great Lake states, you would first have to develop a list of objectives from that chapter. The list would have to include all of the topics, terms, and concepts that were covered in that chapter.

When a student earns 75% on a criterion-referenced test, it is expected to mean more than that the student correctly answered 75% of the questions. The test score should let the teacher know that the student has successfully mastered 75% of the objectives from the material that was covered by the test. In order to do that teachers have to develop a list of objectives and then make sure that each objective is being measured by, at least, one item on the test. It is only when we have completed each of these steps that we are using a criterion-referenced interpretation.

A Comparison of Norm-Referenced and Criterion-Referenced Interpretations

Today, teachers use both norm-referenced and criterion-referenced interpretations most of the time. Each approach has advantages and disadvantages. Let's look at some of them.

Advantages of the Norm-Referenced Approach

One advantage of the norm-referenced approach is that students are not disadvantaged (at least, in terms of their grades) by poor instruction. If everyone in the class gets a low score because the material was not taught well, students can still earn reasonable grades. In a similar manner, with the norm-referenced approach students are not really disadvantaged if the test was more difficult than the teacher had intended, because the grades are typically adjusted by the mean score for the class. So, if a student earned a low test score but did well in comparison to the other students who took the test, then that low test score could still translate to a relatively high letter grade.

Disadvantages of the Norm-Referenced Approach

One disadvantage of the norm-referenced approach is that it is based on competition. No matter how well a student does with material in class, only the brightest, most talented, and hardest-working students will earn good grades. If you are not one of the brightest students in the class and you work hard and master the material well, you still can be destined to earn a mediocre grade. For example, I have known some high school students who elected not to take a higher-level class, such as advanced mathematics, in their senior year simply because they knew that they would be in a class of highly talented students. They feared that, in comparison to their highly talented classmates, they would not fare well and would earn a lower grade. Instead, they chose to enroll in a less challenging class where they would be one of the brightest students and would earn a high grade.

A second disadvantage of the norm-referenced approach is that a test grade frequently does not tell others what the student has learned. If one of your high school students earned an A in Algebra I the previous year in a class where grades were based on norm-referenced interpretations, you really do not know how well the student mastered algebra. The only thing that you really know is that the student was one of the better students in the class. If the class consisted of only students with poor math skills, or if the level of instruction was poor, then the student may have earned an A, but still learned very little algebra.

Advantages of the Criterion-Referenced Approach

One of the clear advantages of the criterion-referenced approach is that when it is applied appropriately a student's grade is highly reflective of the student's level of mastery of the material. In some instances the approach tells us exactly which skills the student has mastered. In fact, it is this characteristic of criterion-referenced interpretations that has made them so popular over the past 40 years.

Another advantage is that students typically have a better sense of how they are performing in the class. Students are better able to monitor their own progress in a class. Am I doing well or do I need to get some extra help with this material?

A third advantage is that this approach is non-competitive. One student's grade on the test is not dependent on how other students performed. Over the past 30 years, American education has moved away from using primarily competitive classroom practices to using more and more cooperative practices. Criterion-referenced approaches fit in nicely with cooperative classroom practices.

Disadvantages of the Criterion-Referenced Approach

One disadvantage of this approach is that students may be penalized for poor instruction. If the teacher does an inadequate job of teaching the material, then the students are likely to perform poorly on the test, which will result in low grades. In a similar manner, if a teacher designs a test that is more difficult than was intended, the students may perform poorly on the test and earn low grades. From the students' perspective, these are disadvantages. However, some educators might argue that these issues make teachers more accountable to provide good instruction and appropriate tests.

Another disadvantage of the criterion-referenced approach is that it cannot be applied to all testing situations. If a teacher is giving a test on a single chapter with a relatively small number of objectives, then it is possible to design a criterion-referenced test. However, on a unit exam that covers four chapters, it is likely that there are too many objectives to develop a test that covers all of them in the time the teacher has scheduled for the test. In that case the only reasonable approach is to develop a test that samples the objectives and apply a norm-referenced interpretation.

There is still another disadvantage of the criterion-reference approach. Many teachers believe that they are using a criterion-referenced approach simply because they have developed some pre-set standards. However, they fail to develop a list of objectives and they fail to design their test to make certain that they have appropriately covered those objectives on the test. Without that second step these teachers frequently have developed tests that are difficult to interpret appropriately.

CHOOSING A FRAME OF REFERENCE FOR ASSESSMENT

As I mentioned earlier, psychometricians do not typically recommend that teachers use either ability-referenced or growth-referenced interpretations because of the problems involved with each. However, there are times when working individually with students when each of these approaches might be useful. Sometimes students incorrectly believe that they are incapable of performing a task because they think that they lack the necessary skills. Using the pep-talk approach, "I know that you can do it," is rarely all that helpful by itself. However, teachers can also work with students to help them develop the necessary skills that they need to be successful with that task. Once the students are more successful, they frequently change their opinion about their abilities and show continued improvement.

The growth-referenced approach can also sometimes work when teachers meet individually with students. Showing Susie two samples of her work, one from earlier in the year as well as a recent sample, can sometimes demonstrate to Susie how much her skills have improved. Both approaches can be used to help motivate students who are convinced that they are not progressing.

Clearly, the norm-referenced and the criterion-referenced approaches are most often recommended by psychometricians. Criterion-referenced interpretations are the most appropriate in situations where there is a clear set of information or objectives that the test is expected to cover. With many types of assessment that is the case. When you give a preliminary or placement assessment you have a certain set of skills that you wish to measure. Are your students already proficient with these skills, or are there skills that you will need to review? Clearly, criterion-referenced interpretations will work best here since they are skill specific. When you use diagnostic assessments, you are asking, "With which particular skills is this student having difficulty?" Again, this calls for criterion-referenced interpretations. When you use a formative assessment, you are asking, "Are my students displaying enough progress with these particular skills that we are ready to move ahead to the next topic?" Once again, this calls for the criterion-referenced approach: You want to identify particular skills. So, as you can see, criterion-referenced interpretations are the most appropriate with preliminary assessments, with diagnostic assessments, and with formative assessments. In each case, you are attempting to identify student proficiency with particular skills, something that the criterion-referenced approach does well.

That leaves us with summative assessments, such as chapter, unit, and final exams. Here we have more choices. If the test is covering a single chapter with relatively few objectives, you can often design a test that will measure each objective. In that case, using a criterion-referenced approach seems most appropriate. However, if the test is covering a relatively large number of objectives, then it is frequently impossible to develop test items to measure each objective and expect students to be able to complete the test in the time available. In that case, you have to choose a sample of objectives from each chapter and develop test items for them. Once you have done that, it is no longer reasonable to apply a criterion-referenced interpretation. You must then use a norm-referenced approach, which works quite well in such situations.

There are other situations where we also might want to use a norm-referenced approach. Let's say, for example, that Laurel Green High School found that it could offer an advanced math class to 20 students and knew in advance that there would be more students wanting the class than could be accommodated. One reasonable approach to choosing which students to accept would be to use their scores from a recent math achievement test. In that case they would choose the 20 students with the highest math achievement scores, believing that they are the students who will benefit the most from the class. Any time you would need to rank students, a norm-referenced approach is appropriate.

CHARACTERISTICS OF NORM-REFERENCED AND CRITERION-REFERENCED TESTS

Should we design norm-referenced and criterion-referenced tests in the same way? The typical answer to that question is, "no." There are many tests that are well designed and that could be appropriately interpreted with either approach. However, if you have one of these two approaches in mind when designing a test, there are several characteristics that you should keep in mind.

Item Difficulty

When using a norm-referenced approach you essentially want to rank students from those who know the material well to those who are less prepared. You want to differentiate among the students, to separate the best prepared and the least prepared students. In order to do this reliably you need to see a large standard deviation, a large spread among the scores. If all of the scores bunch up together, then it is difficult to tell whether differences in student performance on the test are because of differences in the various students' skills or as a result of simple measurement error. However, if the scores are spread out further, you can feel more confident that the test is truly differentiating among the students based on their skills.

The amount of spread that you obtain in the test scores is often related to the average item difficulty. For example, if all of the items are relatively easy, most students will do quite well on the test and there will be little spread in the scores. The same will happen if the items are all very difficult. Most students will do poorly on the test and, again, there will be little spread in the scores. You can frequently obtain the greatest spread of scores (the largest standard deviation) when most of the items are of moderate difficulty. Therefore, when using a norm-referenced interpretation, you frequently want the majority of your test items to be moderately difficult. In that case you will obtain a large standard deviation and will be able to differentiate reliably among students. However, if you plan to use a criterion-referenced interpretation, then you do not need to differentiate among the students and item difficulty is somewhat less important.

Number of Items on the Test

In general it is a good idea to use as many test items as time allows. Longer tests (those with more items) are almost always more reliable [...] and have greater content-related evidence of validity [...]. That being said, we can now look at differences between norm-referenced and criterion-referenced tests.

The number of items needed on a criterion-referenced test is dependent primarily on the number of objectives covered by the test. You need at least one item for each objective. With a fairly large number of objectives, the test will typically have only one item for each objective. Time restrictions will simply not allow for more items. However, having only one item per objective is not the ideal situation as greater reliability and accuracy can be achieved by having more than one item for each objective. If the number of objectives is relatively small, then it might be feasible to have two or more items for each objective.

With norm-referenced interpretations there are no hard and fast rules for the number of items required. In this case you will be sampling objectives to cover on the test and will not need to measure each and every objective. You could, theoretically get away with fewer items. However, it is best to include as many items as time will allow.

Although many test items can be used appropriately on either norm-referenced or criterion-referenced tests, there are times when teachers might want to design items specifically based on the type of interpretation they plan to use. When planning on the number of items to use on the test they also may want to keep in mind how the test scores will be interpreted.

SUMMARY

In this chapter we identified four approaches that teachers use to interpret test scores. With ability-referenced interpretations, teachers compare a student's test score with some estimate of how well the student should be able to perform. With growth-referenced interpretations, the teacher compares a student's performance on a particular skill with how that student performed earlier in the year. With norm-referenced interpretations, the teacher compares the student's performance on a test with the performance of some norm group, often the other members of the class. Finally, with criterion-referenced interpretations, the teacher compares the student's performance on a test with some pre-set standards. Since there are problems associated with the first two approaches, they are not used very often. Norm-referenced and criterion-referenced interpretations, however, are more commonly used for classroom tests.

There are differences between norm-referenced and criterion-referenced interpretations so that in certain situations one has advantages over the other. The two approaches also have different requirements in terms of item difficulty and the number of items required.

EXERCISES

1 Name the type of reference being used in the following examples. Choose from *ability*, *growth*, *norm*, or *criterion*.

   a. Amanda's parents were concerned about how well she was doing in biology class. When they contacted her teacher, Mr. Hower, he reported that she had done better than 70% of her classmates on their last exam.
   b. Mrs. Pierce gave her students a pre-test before she began teaching a unit on fractions. Two weeks later, at the end of the unit, she had given her students the same test to see what they had learned.
   c. Mr. Lin has a rule in his classroom that every student must score at least 90% on a spelling test or the student has to retake it the next day.
   d. Because her students learned about the planets last year, Miss Gates expects them to do very well on the solar system exam.
   e. In order to be placed in the Advanced Reading Program at North Hills Elementary, a student must score 170 or better out of 200 on the placement test.
   f. Carlos scored 552 on the math section of the SATs. With that score, he is placed in the 58th percentile, which means that he did better than 58% of all the students who took the SATs at that time.

2 Write your own example of each type of interpretation for a grade or content area that you plan to teach:

a. Ability-referenced
b. Growth-referenced
c. Norm-referenced
d. Criterion-referenced

SPOTLIGHT ON THE CLASSROOM

Mrs. Neville is a reading teacher at Blue Ridge elementary school. At the beginning of the year, she gave her students a test that contained questions focused on various topics that she planned to cover during the year. She wanted to see what her students already knew. She also wanted to compare the reading levels of each of her students for grouping purposes.

In January, Mrs. Neville feels that she needs to start to focus on preparing her students for the upcoming California Achievement Tests (CATs), which are given in this particular school district. She gives her students the same test that they had taken at the beginning of the year to see what they have learned in the past four months. She looks at each student's test individually and compares the two scores to see how much each has learned and the areas that she needs to review.

Mrs. Neville designed a test that is similar to the CATs. Two weeks before the students were scheduled to take the CATs, she gave her test to the students. She calculated the scores in much the same way that the CATs are scored. She identified areas where students scored poorly and emphasized instruction in those particular areas. When it was time for Mrs. Neville's 3rd-graders to take the CATs, she felt they were fully prepared. After all of this practice, she expects her students to do very well.

Name and discuss the frames of reference that Mrs. Neville used in her classroom.

STUDY TIPS

Time Management

For many college students good time-management skills are as essential for success as they are for most adults in demanding jobs. If you know how to manage your time well, you can typically avoid many last-minute rush jobs or that sinking feeling in your gut when you go in to take a test knowing that you did not allow yourself enough time for adequate preparation. However, good time management requires planning. Because many college students work either part time or full time and have family responsibilities, it can be especially challenging to develop a good time-management plan.

Dianna Van Blerkom (2009) recommends that a good first step in developing a time-management plan is to develop a *fixed commitment calendar*. This is an hour-by-hour calendar for one week which begins when you first wake in the morning and continues until you go to sleep. You begin by filling in all of your fixed commitments, which include meal times, classes, work time, exercise time, and so on. Include any activity that you perform at the same time daily or weekly. If you are uncertain about how you use your time, you could keep a *time log* for a week. In this you record all of your activities on an hour-by-hour basis for a full week. That should help you further complete your fixed commitment calendar which, once completed, should tell you how much time you have left for study and leisure.

A second step is to determine how much study time you actually need. For many college courses students typically need two to three hours out of class for every hour that they are in class. A highly demanding class may require more time, whereas a less demanding class could require somewhat less. During weeks when you are preparing for exams or working on class projects, you will often require extra time. In most instances college students can find enough time for study and preparation. Frequently, when students run out of time, it is because they are spending too much time on leisure activities. Students sometimes discover from their time logs that they are spending an inordinate amount of time playing video (or computer) games, watching TV, or just hanging out with their friends. Of course, some college students work full time and still try to carry a full load in school. That type of commitment requires a great deal of discipline.

Another important time-management skill involves keeping up with your work. That means keeping up with reading your textbook and other out-of-class assignments. If you tend to procrastinate, put off tasks that need to be completed, you have to work especially hard to stay on schedule. One of the best ways to avoid procrastination is to begin a task as soon as it is assigned. It also helps to have daily and weekly *to-do lists*, which are especially effective when you have multiple tasks to complete in a limited time. For complex tasks it often helps to break them down into a number of smaller steps, each of which is less daunting. Once you have filled out your to-do list, it can frequently reduce that debilitating anxiety many people experience when faced with seemingly insurmountable tasks. That anxiety can actually cause people to delay starting the tasks, but, once the to-do list is written, that anxiety is often greatly reduced and you can immediately complete some of the tasks.

There are many time-management skills that can make your college experience more positive. You can get your work done, obtain better grades, and still have adequate time for leisure. If you are looking for additional time-management help, you could consult your college learning center.[2]

NOTES

1 In 2005 Pennsylvania introduced a more flexible pass/fail criterion. A college student can pass several of the basic skills tests (reading, writing, and mathematics) with a somewhat lower individual test score as long as her or his composite score from the three tests meets a certain criterion.

2 This material has been adapted from Van Blerkom (2009).

REFERENCES

Stevenson, H. W., Stigler, J. W., & Lee, S. (1986). Achievement in mathematics. In H. Stevenson, H. Azuma, &K. Hakuta (Eds.), *Child development and education in Japan* (pp. 201–216), New York: Freeman.

Van Blerkom, D. L. (2009). *College study skills: Becoming a strategic learner* (6th ed.). Belmont, CA: Wadsworth.

Reading 4.2

FORMAL VS. INFORMAL ASSESSMENT

Malcolm L. van Blerkom

Generally, when you think about assessment in the classroom, I would guess that tests and quizzes are probably the first things that come to mind. Tests (both standardized and teacher-made), quizzes, and other similar devices are referred to as *formal* assessment devices. With formal assessment you, the teacher, are able to complete the assessment in a relatively standardized manner and are able to control many aspects of the process. With formal assessment all students are given the same questions to answer, typically have exactly the same amount of time to respond, and complete the device in a relatively consistent fashion. Formal assessment is important in the classroom; it is not, however, the only type of assessment that we use.

A much more common type of assessment is what we refer to as *informal* assessment. It involves the many observations that we make about students and the many questions that we ask students throughout the day. It may involve having a child read aloud and attempting to note and remember which reading processes are still giving this child problems. It could involve having children go to the board to work out math problems and looking for which students are performing well and which students appear confused. It can involve the many questions we ask students throughout the day and noting how each student responded to those questions. Many times we are using informal assessment to determine if our students are displaying competency with the skills which we have been teaching. Are we ready to move on to the next topic? Clearly, we do a great deal of informal assessment in the classroom. In fact, I have heard educators estimate that perhaps as much as 90% of the assessment that we do in the classroom is informal assessment, especially in the early grades.

CLASSROOM ASSESSMENT

Preliminary or Placement Assessment

We use assessment in the classroom to serve a variety of purposes. Sometimes we use *preliminary* assessment. Sometimes, at the beginning of the school year we are ready to move ahead with our curriculum, but are uncertain as to how well the students are prepared. Let's say that a 5th-grade teacher has learned from previous experiences that she cannot always begin the school year with the prescribed 5th-grade math curriculum. She has found that sometimes her students came from classes in the previous year where the 4th-grade teacher did not complete the prescribed materials or that some students forgot a great deal over the summer. Therefore, within the first two weeks of the school year, our 5th-grade teacher gives her students a test of the prerequisite math skills, the skills that they need to be prepared to master the new skills that she plans to teach them. If her students perform well on this preliminary test then she knows that she can safely move ahead with the prescribed lessons. However, the test is likely to reveal that there are a few skills that she will need to teach first, or at least review, before she can begin the prescribed curriculum.

A variation of this is the *placement* assessment. Placement assessments are very commonly used in colleges where incoming freshmen are often given placement assessments in a variety of subjects. For example, many colleges require students to take one or more math courses and frequently require proficiency with basic algebra. Therefore, they give mathematics placement tests to their incoming freshmen. If students' scores are high enough, they are frequently exempt from taking the freshman algebra class and can take a higher-level mathematics class. These placement tests are especially important in fields like science and engineering where students are typically expected to take a calculus sequence. Are students ready to take Calculus I or do they need to take the Pre-Calculus course first? Placement tests are also used in other areas such as English and foreign languages, where sequenced courses are often available. Placement tests are also used in elementary schools, middle schools, and high schools. Sometimes they are used at the end of the school year to help determine next year's placements.

Definition

Preliminary or **placement** assessments are assessments performed within the first two weeks of the semester that are designed to measure students' prerequisite skills.

Diagnostic Assessment

A second type of assessment is the *diagnostic* assessment. We use diagnostic assessments when we recognize that students are not performing well, but are uncertain of the exact problem. For example, Ms. Kindya, a 2nd-grade teacher, notices that Tim is struggling with reading, yet she is unable to identify clearly which skills are giving Tim the greatest problems. Luckily, Ms. Kindya is familiar with and skilled at using a variety of reading diagnostic tests. She chooses the test that appears most appropriate, administers it, and is able to identify the specific skills that are giving Tim trouble. She now can provide Tim with the specific help that he needs. We are likely to use diagnostic tests most frequently in the areas of elementary reading and mathematics. However, we also have diagnostic tests in a number of other fields.

Although some diagnostic tests are designed for classroom teachers to administer, others require considerable training. In those cases, you may need to call on a specialist, like a reading specialist, a speech-language pathologist, or a school psychologist, to administer such tests.

Definition

A **diagnostic** assessment is any type of assessment used to identify an individual student's learning decrements.

Formative Assessment

The third type of assessment that we use is known as *formative* assessment. Formative assessment is any type of assessment device that we use while an instructional unit is in progress. It is used primarily to give the teacher feedback on how the unit is progressing. Let's say that you are a 5th-grade teacher and are teaching a six-week unit on adding and subtracting fractions. You might recall that in order to add or subtract fractions, you must first find a common denominator. For example, to complete this problem 1/2 + 1/4 you must convert the 1/2 to 2/4. Therefore, the first two to three weeks of the unit are dedicated to teaching the students this prerequisite skill, finding the common denominator. Before you move into the part of the unit where the students actually learn to add and subtract fractions, you want to make certain that they are doing very well with this prerequisite skill. Therefore, at the end of week two you give your students a pop quiz that requires them to complete 10 common denominator problems.

Over the weekend you score the quizzes. In this case, you are using the quizzes primarily as a feedback device for yourself. You want to find out if the class is showing sufficient proficiency with this skill to move to the next step? If they performed well enough on the quiz, then you can move to the next step early next week. If, however, they are still not performing well with the task, it means that you need to spend several more days working with common denominators before you can move on. Formative assessments are designed to give you, the teacher, feedback. Therefore, most psychometricians argue that formative assessments should not be used to give students grades. Others, however, claim that unless assessments are graded students may not take assignments seriously.

Besides pop quizzes, there are many other techniques that can be used for formative assessment. If we continue with the math instruction example, some other techniques that could be used would include having the students perform seat work while you walk around the room looking at their work. You could also have them come up to the board and solve problems there. You could play a game where they have to find the common denominator. There are a large variety of techniques that you can use to gauge students' progress. Many formative assessment techniques involve informal assessment.

> **Definition**
> **Formative** assessment is any type of assessment device that we use while an instructional unit is in progress and is used primarily to give the teacher feedback on how the unit is progressing.

Summative Assessment

The fourth type of assessment is *summative* assessment. Summative assessments are performed at the end of chapters or units to determine the students' level of competency with the material and to assign grades. Many of you may have assumed that summative assessment was the only type of assessment used in school. It is the type that we probably recall most often and with which we have had the most experience. Not all teachers use preliminary or diagnostic assessment. Even when our teachers were using formative assessment we, as students, were not always aware of why they were doing it. However, we were typically keenly aware of summative assessment.

> **Definition**
> **Summative** assessments are performed at the end of chapters or units to determine the students' level of competency with the material and to assign grades.

All four types of assessment play important roles in the classroom. However, in many classrooms formative assessment is the most common type of assessment, as it should be. Elementary school teachers tend to use formative assessment often, especially with the youngest students. As students move into the higher grades, however, there is a tendency to use it less often, which may be a mistake. Many college instructors hardly use it at all. At times, after working with my class on a difficult topic, I find myself asking them this question, "Is that clear to everyone?" As they nod their heads, I feel satisfied that they understand the material and move on to the next topic. I often think that I've completed an effective formative assessment. In the last few years, I have come to recognize that when I do that, I am using a very ineffective approach. How many times, in your experiences, has someone asked you a question like that after teaching you something fairly complex? You may have responded in the affirmative, only to discover later, when you tried to apply the new learning, that you really did not understand. It is much more effective to actually have students try out the new learning to assess if they can use it correctly.

MAXIMUM vs. TYPICAL PERFORMANCE MEASURES

When we measure some skills and abilities we want to assess how well our students can perform. We want to assess them at their best. However, when measuring other types of traits, we are more interested in students' typical behaviors. Classroom tests designed to measure a student's level of competence with some skill should be geared toward measuring the student's maximum performance. We want to assess them doing as well as they can do. However, personality tests or career interest inventories and similar measures work best when they measure a student's typical behavior or interest.

There are differences in how we design the most effective maximum performance measures compared to how we design effective typical performance measures. If this text were designed to be used in a course on psychological measurement, we would spell out those differences. However, since this text is designed primarily for classroom teachers, that is not really necessary. The tests and other measures that classroom teachers use will, almost exclusively, be maximum performance measures. That will be our focus in this book.

Definitions

Maximum performance measures are used when we want to assess how well our students can perform.

Typical performance measures are designed to assess a student's most common, or typical, behavior.

USES OF MEASUREMENT

Since this chapter is introducing the section in this part on measurement, this might be the appropriate place to let you know what to expect in the remaining chapters. Chapter 2, Frames of Reference, will introduce you to the various ways in which we can interpret test scores. Chapter 3, Developing Objectives, covers the various approaches that educators have used to establish goals and objects. We use objectives to help you decide what you will teach and what you will test. Chapter 4, Reliability, will introduce you to the concept of reliability, consistency within tests. What is reliability? How do we measure it? How can we develop tests that are reliable? Chapter 5, Validity, deals with the issue of whether a test is actually measuring what we think it should be measuring.

Chapters 6 through 11 deal with building effective classroom tests. Chapter 6 discusses short-answer and completion items. Chapter 7 deals with essay items. Chapter 8 deals with multiple-choice items. Chapter 9 deals with true-false items and other alternative choice variations. Finally, Chapter 10 discusses how to assemble the individual items into a test. Chapter 11 is a follow-up on the earlier chapters on classroom testing. It deals with how to analyze your tests to assess if they are performing as you had hoped they would.

Chapters 12, 13, and 14 deal with alternative forms of assessment. Chapter 12 delves into informal assessment, which was briefly introduced in this chapter. Chapter 13 deals with performance assessments, which we use when we need to rate a student's actual performance as a judge would do when rating an athletic competition such as gymnastics. Chapter 14 discusses how to use portfolios effectively.

The last three chapters deal with other important topics. Chapter 15 deals with teaching students to be effective test takers. Chapter 16 describes the various types of standardized test you are likely to encounter in schools. Finally, Chapter 17 deals with many alternative ways to report test scores and why many of these are, at times, better than reporting raw scores.

SUMMARY

In this chapter we examined the various roles that measurement plays in the classroom. We also differentiated among three commonly used terms: assessment, measurement, and evaluation. We differentiated between formal assessment and informal assessment. Classroom tests, quizzes, and similar devices are frequently referred to as formal assessment, whereas teacher questions, observations, and many other techniques are often referred to as informal assessment. We differentiated among four ways that we use assessment. We use preliminary or placement assessment to identify students' prerequisite skills. We use diagnostic assessment when students are experiencing learning problems. We use formative assessment to judge how well our instruction is proceeding. We use summative assessment to judge our students' competence with the material that they have been taught. Finally, we differentiated between maximum and typical performance measures. In the classroom we typically use maximum performance measures because we want to assess how well our students can perform the new skills that they have learned. In other settings, especially when examining concepts like personality, we use typical performance measures which will identify the individual's typical, or usual, behavior.

HOW DO VARIOUS TYPES OF ASSESSMENTS APPLY TO MY PROFESSION?

As described above, there are multiple types of assessments, each with different characteristics and purposes. As an educator, whether a classroom teacher or educational leader, what you do with the results of any assessment is as important as the assessment itself. The decisions you make about student learning can have a tremendous impact on future successes and/or failures. There are significant consequences associated with the results of assessments. Therefore, it is imperative that you ensure you make decisions that align with the purpose of the assessment. You would not want to make consequential decisions about student learning from formative assessments. Alternatively, the reality of making decisions about student learning needs to come from somewhere. Results on classroom assessments can be one of the best indicators of what students know and can do. Be careful, though: You want to ensure your assessments meet the needs of each individual student. Thus, you may assess student knowledge on one topic differently for each student. Assessing student learning should be flexible to meet the needs of each individual student. For example, one student may perform better orally and another in written format. It does not mean that one student knows more than another. It simply means that as a professional educator, it is your responsibility to meet the needs of all students.

CHAPTER 5

Test Formats and Guidelines for Developing Assessments

I develop my own assessments: What is wrong with that?

AS YOU READ

As you read the chapter, think about the following:

- How do **you** define or explain test format?
- What is **your** experience with various test formats?

WHAT ARE THE GENERAL PRINCIPLES OF TEST FORMATS AND DEVELOPING TESTS?

There are multiple types of tests as described in Chapter 4. In addition, regardless of type of test, there are various formats for tests. This chapter will review various formats for tests, the pros and cons of each, and the appropriateness of each type based on the purpose of the test.

Test Blueprints: What Are They and Why Do I Need One?

When you develop any test, you should start with a test blueprint. A test blueprint is a framework or map that is developed to ensure that all aspects of the curriculum and constructs (or domains) are assessed. Test blueprints are often used to improve the evidence of content and possibly even construct validity of a test.

Let's walk through a typical example of classroom assessment. You, as an educator, want to conclude what students know about the U.S. Constitution. Thus, you plan a 4-week unit about various aspects of the U.S. Constitution. You identify which components you plan to teach during multiple lessons during the 4-week unit plan. For example, you decide to teach (obviously what you decide is based on appropriate standards that drive the curriculum) about the development of the Constitution, the original participants, the amendment process, and the amendments themselves. You spend a great deal of time preparing each lesson that you will teach over the next 4 weeks. Throughout the unit, you utilize multiple formative assessments that provide you feedback about student learning throughout the unit. Using these results, you make some slight changes to the timeframe for the unit, some instructional strategies, and you tweak your lessons as a result. This is a fairly typical classroom scenario that you do on a regular basis—it is actually probably second nature.

So, where does a test blueprint come into the scenario? The answer is during the planning phase. Just as you spent hours thinking about the curriculum and instructional strategies, you should spend as much time on planning for the assessment of student learning. As you plan, you should develop a map or a framework that will guide the content for the lessons and the assessment. So, based on the concepts you decided to teach, you write those on paper (there are multiple examples and templates available online for you to use to guide your test blueprint). You list the main concepts for the unit. These main concepts guide what you teach and what you want the students to know at the end of the unit. These should align—what you teach is what you assess! Now, you have the main concepts listed on paper. Next, you write sub-concepts. For example, you want them to know about the amendment process. The amendment process is the main concept. But, what specifically do you want them to know about the amendment process? Who is involved? How the Constitution can be amended? You must know exactly what you want the students to know about the amendment process to teach them about it and then assess their knowledge. The specifics you want them to know about the main concepts is the next items you write on the test blueprint. Finally, you write an appropriate number of questions related to each of the sub-concepts. Write those on paper as well. The final step in this process is for you to get feedback from one or more educators of the same discipline as well as an assessment expert. You are trying to ensure that you teach the content thoroughly and that you assess what you teach. This process does take some time and it is extremely important that you take the time on the front end. This helps to ensure that you have evidence of validity, so you can make appropriate decisions based on the results of the end-of-unit assessment.

Multiple Response Tests: What Are They? Pros and Cons

Multiple response tests are most commonly known as multiple-choice tests. These are tests that include two components, the stem and the leaves. Other names for stem and leaves are the question and the responses. Most educators (and students) are very familiar with multiple response test questions. Although you may already be familiar, there are some important characteristics and guidelines that should be used when creating tests and/or determining the quality of existing tests. The following includes some guidelines if you decide to develop your own multiple response tests.

Determine the number of possible answers (responses/leaves) you want for each item. Whatever you decide should be consistent throughout the entire test. If some items have four possible responses and others have three possible responses, you increase the change of error on the test that reduces the evidence of reliability. Keep in mind, the number of possible responses is directly related to the probability of students obtaining the correct response as a result of guessing. This means that if the item has three possible responses, then there is a 33% chance of students obtaining the correct answer by guessing alone. Four possible responses results in a 25% chance of guessing the correct response. Five possible responses results in a 20% chance of guessing the correct response. Once you have decided the number of possible responses for each test item, you can begin writing items. Four possible responses are the most common. However, three is typically enough to use in classroom settings. Given the difficult nature of developing four plausible answer choices, many testing experts agree that three are advisable.

Just as the number of possible responses should be the same throughout the test, the format for the stems should be consistent. The more you change the format of the items, the more chance of error simply because you can confuse students. The best advice is to keep items simple and only include the necessary verbiage. Do not include words that are not absolutely necessary for the test item. Also, decide if you will write questions or statements for each item. Whatever you decide, stay consistent throughout the entire test.

The following are things to avoid in multiple response tests:

- **Avoid negative wording.** Negative wording can confuse the examinee. Sometimes, these types of questions create double negatives that are very confusing.
- **Avoid the phrasing "None of the above" and "All of the above."** These responses can easily confuse the examinee. Ask exactly what it is that you want to know and then provide only one correct response.
- **Avoid the phrasing "Choose the best response."** The term "best" is inherently an opinion. Yes, sometimes one answer is better than the other; however, it is best simply to provide one correct answer—not one that is better than the other.
- **Avoid giving clues in the responses.** If the question asks about specific dates, then each answer choice should include dates. If one does not include a date, you are giving a clue that the particular answer choice is not viable.
- **Avoid incorrect grammar in the responses.** If the item asks for a plural response, then all possible answer choices should be written in plural form. Another common error is the use of "a" and "an." Make sure that the question and the answer choices are grammatically correct and do not give clues by the use of "a" versus "an."

Remember, you are trying to reduce error in the test so you can determine what students know and can do. When you include things listed above, you decrease the likelihood of reliability of results of the test.

Pros of Multiple Response Test Items

A common argument for using multiple response tests is the amount of time they take to grade. Yes, as an educator, your time is valuable. It is understandable that how much time it takes to grade an assessment is no doubt a consideration. However, make sure that the format of the test matches what it is you want the students to demonstrate. If you are assessing facts, multiple response tests are commonly used. Conversely, if you are measuring performance, multiple response tests may not align to the information you are trying to collect.

Cons of Multiple Response Test Items

Multiple response tests cannot assess everything as indicated in the pros paragraph. Knowledge may be assessed using multiple response tests. However, constructs such as performance and skills and questions that use verbs such as "describe," "discuss," and "explain" are not easily measured using multiple response tests. The usage may be limited in some circumstances. Make sure the format of the assessment aligns to the particular content standard for which you are assessing students. Another drawback to multiple response tests is that they are extremely difficult to write. Writing good test items at first glance appears simple. However, it can be very challenging to follow all the guidelines and avoid all the items listed above.

Dichotomous Response Tests: What Are They? Pros and Cons

Dichotomous response tests are typically tests with two possible answers, such as right/wrong, yes/no, or true/false. These are a subgroup of multiple response tests. Thus, the same guidelines described above apply to dichotomous response tests except there are two possible answer choices instead of multiple answer choices.

Pros of Dichotomous Response Test Items

Similar to multiple response tests, dichotomous response tests are fairly straightforward and quick to grade. This is often the number one factor considered when using dichotomous response tests.

Cons of Dichotomous Response Test Items

Dichotomous response items provide a 50/50 opportunity to guess correctly. Thus, students have a 50% chance of guessing the correct answer although they may not know the content. These results may lead to invalid conclusions about students' knowledge about concepts.

Fill-in-the-Blank Tests: What Are They? Pros and Cons

Fill-in-the-blank test items are commonly used for fact recall or rote memory. These are considered closed-ended questions because you are seeking one- or two-word responses. Some general guidelines include making sure the blanks are the same for each possible answer. If you make the sizes of the blanks differently, you are giving clues to the length of the answer. Also, just as the others already described, you need to make sure the items are grammatically correct without giving clues. For example, you do not want to use the words "a" or "an," which can give clues to the starting letter of the response. Instead, you can use a/an as a possibility.

Pros of Fill-in-the-Blank Test Items

As is the case for other closed-ended items, these take little time to grade because you are seeking specific verbiage in the responses. Also, these items can be more straightforward and less challenging to write.

Cons of Fill-in-the-Blank Test Items

Fill-in-the-blank test items typically only assess basic knowledge and/or rote memory. It is challenging to write items that get the students to think beyond one or two words. Also, you must consider what responses you will accept. Will you only accept the exact word that you are seeking? What if the student uses a synonym of that word? Will you accept that response? The answers you are seeking for demonstration of knowledge are not always crystal clear making this a con of this format of test item.

Open-Ended Tests: What Are They? Pros and Cons

Open-ended test items are most commonly referred to as essay test items. These items are best used when students need to describe or even explain concepts. Open-ended items allow students to incorporate their thoughts, feelings, and attitudes about concepts. These types of questions typically begin with how, why, or explain. When writing open-ended test items, you should always assess only one concept in each question. For example, if you want the students to explain their knowledge of the constitutional amendment process, you should ask them to explain the constitutional amendment process. Do not ask them generally about constitutional amendments that may leave them guessing about what it is you want them to explain. The wording of these questions is extremely important. A key element is to guide the students in the expected responses without leading them to their answers.

Pros of Open-Ended Test Items

As stated above, open-ended test items allow students the opportunity to include their thoughts, feelings, and attitude about various concepts. This often provides the teacher with an in-depth understanding of what students know and think about the concept. These details can help guide the teacher to learn more about his/her students and the depth of their knowledge.

Cons of Open-Ended Test Items

Open-ended test items allow the opportunity for students to explain (written or orally) what they know about concepts. However, the open-ended nature does not provide much guidance for expected results. Thus, students may include all types of responses and details that do not directly relate to the concept that is being assessed. Also, these tests can be very time consuming to grade. Finally, when grading open-ended test items, you must decide what you will consider. Will you consider grammar in their response? If so, how much weight will you give grammar? Will you consider the length of the response? If so, how much weight will you give the length? There are multiple considerations when grading open-ended test items. Often, rubrics are recommended as a guide for students' responses. Also, rubrics can help the grader with consistency across all students.

WHAT DOES THE LITERATURE SAY ABOUT TEST FORMATS AND DEVELOPING TESTS?

Reading 5.1

PRODUCING AND ADMINISTERING TESTS

Malcolm L. van Blerkom

INTRODUCTION

Now that you have learned how to prepare good short-answer, essay, multiple-choice, and true–false items, you need to learn how to plan, produce, and administer a classroom test. In this chapter we will discuss how a teacher goes about planning a test, choosing which item formats to use, deciding on the difficulty level of the items, and producing the test. We will also discuss how to set up an appropriate testing environment and how to administer a test.

DESIGNING A TEST

Many new teachers develop classroom tests without spending enough time planning what they want to do. With experience, they soon learn that planning is critical to developing high-quality tests. Teachers need to spell out the purpose of the test, the material that will be covered, the types of item that will be used, the difficulty level of the items, and the time that is available for the test—just to name a few of the steps that must be taken.

Defining the Purpose of the Test

You may recall from Chapter 2 that one of the first things a teacher must do in designing a test is to define the purpose of the test—decide how the test scores are going to be used. Is the test a formative assessment designed to let you know how the students are progressing in developing a particular skill? Or is the test a summative assessment given at the end of a chapter or unit to allow you to assign grades based on the students' performance?

Let's say, for sake of discussion, that you are developing a summative assessment, a unit test at the end of a three-chapter unit. The test scores will form a major part of the grades that the students will be assigned for the marking period, which means that the test will be covering a fairly large set of objectives. If the number of objectives is small enough, the test will be able to cover each objective and you can plan to use a criterion-referenced assessment—an assessment that covers all of the objectives from the content domain. However, if the number of objectives is too large, then you will have to sample the objectives on the test and use a norm-referenced interpretation—an assessment that samples objectives from the content domain.

Choosing the Types of Items to Use

You must choose the types of item format that you will use. Will the test be all short-answer items? Will the test contain items from several formats? This choice is based largely on the type of material that will be covered. If the objectives are primarily at the knowledge and comprehension level, then short answer, multiple choice, and true-false items are all appropriate. If there are higher-level cognitive objectives involved, then multiple choice or brief essays are more appropriate. Many classroom teachers tend to use two or more item formats on many of their tests. How many of your teachers did?

Choosing the Number of Items to Be Used

Let's say that you have chosen to use only short-answer items. How many should you use? There are actually a number of issues to consider before you choose the number of items to use. One of the first issues must be time. If you are working with classes that last only 47 minutes, then you must be able to administer the test within 47 minutes. Those 47 minutes must also include time to take roll (if that is required) and deal with any other class business, pass out the testing materials, go over the instructions with the students, and collect the testing materials. If you like to review a test on the same day that it is administered, then you have to allow time for that, too. In this case, you may only be able to give the students 30 minutes to take the test.

Once you decide on a realistic time frame, you have to predict how many test items the students can complete in that time. Students can typically complete three to four simple short-answer items per minute if the items require one-word answers. More complex short-answer items can take about twice as long to answer. Computation items can take even longer, depending on the complexity of the computations. Similarly, students can answer three to four simple true–false items per minute. Multiple-choice items frequently take longer. Students can answer simply worded multiple-choice items at the rate of about two per minute, whereas more complex multiple-choice items can easily take a minute or longer to answer. Short essay items should take about 10 minutes to answer. These time estimates are only averages. Some students can take longer with certain types of item.

Choosing the Difficulty Level of the Items

What is the optimum level of difficulty for the items? This is also a complex and, sometimes, still debated issue. There are two primary issues here: The first has to do with the difference between speeded tests and power tests (discussed briefly in Chapter 4).

Speeded Tests vs. Power Tests

Speeded tests are tests with very specific time limits. They are designed to see how quickly students can perform certain simple tasks. With a speeded test there are more items on the test than most students can be expected to complete within the time limit. The items are simple so that students are expected to make few, if any, errors. The score is based on the number of items correctly completed. Speeded tests are most appropriate when measuring manual and clerical skills like typing. They are not appropriate when measuring most cognitive skills. Power tests, on the other hand, do not typically have time limits (although some time limits are necessitated by practical issues). Students are expected to complete all of the items. However, the items are more difficult. With power tests teachers are attempting to see how well the students can use the knowledge and skills that they have learned. In conclusion, speeded tests will typically involve many easy items, whereas power tests will involve fewer items that are moderately difficult.

Criterion-Referenced vs. Norm-Referenced Tests

The second issue has to do with criterion-referenced assessment versus norm-referenced assessment. With criterion-referenced assessment the item difficulty is not a very important issue. Items used in these tests can be anywhere from moderately difficult to fairly easy. A large number of students can earn high scores on a criterion-referenced test if the students are all competent with the skills being measured. Norm-referenced tests, however, are different in that they are designed to rank the students—to differentiate among the students who know the material well from the students who are less well prepared. This differentiation can only be accomplished if the test scores have a relatively large standard deviation—if the test scores are spread out. We want the students who are less prepared to obtain relatively low scores, the students who are moderately prepared to earn higher scores, and the students who are the most prepared to earn the highest scores. If the test items are too easy, most students will get high scores and the test will not differentiate well. If the test items are too difficult, most of the students will get low scores and the test, once again, will not differentiate well. We tend to get the largest standard deviation and the best differentiation when test items are moderately difficult (when about 45% to 85% of the students get the item correct, depending on the type of item). Therefore, we typically want more difficult items when we use a norm-referenced assessment.

Assuring Sufficient Accuracy

Most tests are used to help teachers make decisions about students. In order to make good decisions that are based on valid information, our tests must have a sufficient amount of accuracy. Because this is a somewhat complex issue it can best be demonstrated through an example. Let's say that a course is operated on a pass/fail system and, for simplicity, let's say that the grade in the course is based on one test only. In addition, let's say that a student must score 75% or above on the test to pass the course.

In such a situation four possibilities exist. Some students will be competent with the material and should pass the course, whereas other students will not be competent with the material and should not pass the course. The first possibility is that a student is competent with the material and gets a 75% or above on the test. The student passes the course and a correct decision has been made. The second possibility is that the student is not competent with the material and scores below 75% on the test. The student fails the course, and, again, a correct decision has been made. The third possibility is that the student is competent with the material but scores below 75% on the test. The student fails the course when he or she should have passed and a decision error has been made. This is sometimes referred to as a "false negative." The fourth possibility is that the student is not competent with the material but gets a 75% or above on the test. The student passes the course when he or she should have failed it and, again, a decision error has been made. This is sometimes referred to as a "false positive."

Overall, we would like all of our decisions to be correct decisions. We would like to avoid false positives and false negatives—misclassifications. However, there probably has not yet been a test invented that always makes correct decisions. Even with medical lab tests, it is not unusual to get these errors. Nevertheless, we can reduce these types of error with careful planning.

Why We Get Misclassifications

One reason why we get misclassifications is that we never know a student's ***true*** score—the score a student should receive with a perfect test. Because measurement error exists, students typically receive test scores a few points lower or higher than their true scores. For students whose true score is far from the cut-off score, misclassifications are rare. If a student's true score is 55%, it is very unlikely that he or she will obtain a test score above 75%. The student will probably get a failing grade and a correct decision will be made. On the other hand, if a student's true score is 95%, it is very unlikely that he or she will score below 75%. The student will probably get a passing grade and a correct decision will again be made. However, for students whose true scores are between 70% and 80%, close to the cut-off score of 75%, there is a much greater likelihood that they will score either too high or too low and a decision will be made in error. In such a situation, it might be possible for 30% (or more) of the students whose true score was within five points of the cut-off score to be misclassified.

How to Reduce Misclassifications

One way to reduce these misclassifications is to develop tests with greater reliability. With greater reliability, the difference between true scores and obtained scores is smaller. The simplest way to increase reliability is to have a large number of items on the test. When we have a test that will be used to make important decisions we want as many items on the test as we can possibly fit in. In that case, we can reduce the number of misclassifications.

When important decisions are to be made we want to use tests that are as accurate as possible, and frequently we want to use multiple tests. Let's use a medical example. Does an individual have type 2 diabetes? Recent research suggests that we can reduce many of the negative effects of diabetes if we begin treating it as early as possible. Since type 2 diabetes (also sometimes known as adult onset diabetes) can take 15 or more years to fully develop, we would like to pick it up as early as possible before there is sufficient damage to the patient's pancreas. Therefore, physicians will typically require their patients to have yearly blood tests. Part of those blood tests will be a check for blood glucose level. Let's say that our patient, George, has always had blood glucose levels within the acceptable range. However, this year the test score is five points above the acceptable level, indicating possible type 2 diabetes. Now, this test score could be an error. Perhaps George got up in the middle of the night before the test and had a glass of orange juice when he was supposed to fast. A cautious physician would not be likely to diagnose George as diabetic based solely on this one elevated test score. The physician could order other blood tests or simply have George repeat the test in a month or two. If those later tests fall within the normal range, the physician is likely to assume that the elevated blood test was an error, but will carefully monitor George's blood glucose level in the future. On the other hand, if the other tests confirm the first elevated test, then the physician is likely to diagnose George as having type 2 diabetes and will begin treating him.

The same applies to classroom tests. Because you are making important decisions about students, whenever possible, you should avoid making those decisions on the basis of a single test score. You should always try to obtain corroborating evidence to make certain that you have made the best decision that is possible. You should look at other tests and other indicators of the same skills. Decisions based on multiple test scores tend to be far more accurate than decisions made on a single test score.

PRODUCING A TEST

Now that you have planned the test, you have to prepare the items and produce the test. There are a number of issues to keep in mind when producing a test.

Preparing the Items

Generally, the most time-consuming part of developing a test is preparing the items. By using your list of objectives or your Table of Specifications, you can begin to write the items. Writing all of the test items yourself takes a great deal of time. Today, most textbooks come with a teacher's edition that contains sample tests or test banks, which are frequently very helpful. However, these test banks are sometimes prepared by graduate students hired by the textbook author to produce the test items, and some of these graduate students have had only minimal experience in producing test items. Therefore, many of the test items found in the teacher's edition are potentially problematic. You should go through the test bank to look for possible items. At times the items can be used as written. However, at other times you may encounter a potentially good item that is somewhat problematic as it is currently written. In that case, you can simply start with that problematic test item and edit it to improve it. Since editing items found in teacher's editions often takes less time than writing items from scratch, I encourage you to try that approach.

Ordering the Items

There are no hard and fast rules for the order of items on a test. Research suggests that for most tests, item order has little effect on how students perform on the test. It does not appear to matter if the test begins with a few easy items or a few hard items. Therefore, the order of items should primarily be determined by other practical considerations. For example, it is frequently easier to score a test when items of one format are grouped together. In addition, if you routinely review tests, then it is also helpful to keep items grouped by topic. It is also common to have the more objective items at the beginning of the test and the essay items (if any) at the end. However, in spite of these recommendations, if you find that another way to order the items works best for you and your students, then I encourage you to try it.

Formatting the Test

The two biggest issues in formatting a test are to make the test easy for the students to read and complete, and to make it easy for you to score. Since most of you will be preparing tests on a computer, using a word processing program, formatting will be much easier than it used to be.

There are several things to keep in mind to make the test readable and better for your students. First, do not try to cram too much material onto a single page. Cramming too many items onto a single page can make the material difficult for the students to read. In general, it is best to leave plenty of "white" space. The items need to be clearly prepared so that one item does not blend into another; so, double space between items. In addition, the students should not have to flip pages to complete an item. If an item cannot fit at the bottom of the page, begin it on a new page. Students should not have to flip pages to answer a question based on a diagram, table, map, or chart. The required material should all be on the same page as the questions about it. For younger students, where questions are often brief, some teachers have found that printing the page in a two column format makes the items more readable.

The second issue is to make the test easier for you to score. If you do not use a separate scoring sheet,[1] it can be helpful to format the items in a manner that will ease scoring. Often, the best way to do that is to have the students answer the question in either the left or the right margin next to the item. You can put blanks in the margin to clearly indicate to the students where the answers should be placed. You can then hold your key next to the students' answers and score the items quickly.

Preparing Instructions

Except for the youngest students, all instructions should be provided in writing at the beginning of the test, and when appropriate at the beginning of each new section. The instructions need to be clear, yet concise. The instructions need to describe the number of items and the number of pages on the test. They should include directions on what the students need to do to complete the items and any special instructions (for example, whether they can use a calculator or scratch paper). Instructions should also include the time limits. In the directions before each section, you should also list the point value of each question.

Of course, you should read the instructions to younger students. Do you think that it is necessary to read instructions to older students if the instructions have been written on the test? Many teachers have found that, although they tell their students to read the instructions before starting to take the test, many never do so. Frequently, they are so anxious to get started that they simply skip the instructions. Therefore, I strongly recommend that you read the instructions aloud to the students. It never hurts and, often, it can help.

Proofreading

The last step before printing the test is to proofread it carefully. Take the test yourself as the students would take the test. Doing so sometimes reveals errors that you would easily miss if you were simply reading through the test. Making announcements about typos while the students are taking the test wastes time and distracts some students. Doing so also interferes with the students' concentration.

ADMINISTERING THE TEST

Setting Up an Appropriate Testing Environment

You may recall from Chapter 4 that measurement error can come from three different sources, one of which is environmental effects. A test can only be reliable and valid if it is administered in an appropriate environment. Students need a comfortable, quiet place, free from distractions, to take a test. The room should be well-ventilated and be at a comfortable temperature. There should be few distracting noises or activities going on in the classroom. The room should not be crowded. Students need to feel that they have enough room to work without being on top of one another.

Minimizing Frustration

Teachers also need to minimize student frustration. Frustration often leads to anxiety which tends to have a negative effect on student performance. Testing needs to follow a reasonable schedule. Students need to be given plenty of advanced warning about an upcoming test so that they have sufficient time to prepare. Many teachers provide review sessions the day before the test, which can allow students to judge their level of preparation and gives the students the chance to devote additional time to study, if needed. Although teachers need to stress the importance of tests, it is probably best not to over-emphasize a test's importance. That only serves to heighten student anxiety.

Minimizing Interruptions

It is also important to minimize interruptions during the test. If the test is well prepared and the teacher gives instructions before the students begin the test, there should be no reason for the teacher to have to give additional instructions during the test. What about student questions? In a similar fashion, we do not want students asking questions aloud during the test as that can also interfere with the other students' ability to concentrate. Some teachers, therefore, allow no student questions during a test. I, however, recognize that students sometimes have difficulty interpreting questions even if they are well prepared. Because language is so complex, some students will sometimes interpret a word or phrase to mean something other than what was intended. If the students have a question, they could come to the front of the room to ask for clarification. You can answer the students' questions, in a quiet voice, unless your response would directly provide them with the answer to the question. Some teachers are unhappy with that procedure because they believe that, as students come to the front of the room, they will look at other students' tests. They ask the students instead to raise their hands and then go to them. There are a variety of ways of dealing with student questions that can minimize interruptions.

SUMMARY

A well-developed test requires careful planning. A teacher must decide on the purpose of the test, the type and number of items to be used, and the difficulty level of the items. The teacher must also be sure that the test has sufficient accuracy for the decisions that will be made based on its results. The teacher must prepare the items using objectives or a Table of Specifications, decide on the order of the items, format the test, and prepare instructions. It is also important for the teacher to carefully proofread the test before having it printed. When administering the test, the teacher must set up an appropriate testing environment that allows students to perform at their best. The teacher should also try to minimize both student frustration and interruptions during testing.

EXERCISES

1 Label each of the following as a descriptor for a *power test* (P) or a *speeded test* (S):

a. __________ Contains a large number of items
b. __________ Contains fairly simple items
c. __________ Contains difficult items
d. __________ Time is not a factor
e. __________ Appropriate for measuring manual/clerical skills
f. __________ Contains a specific time limit
g. __________ More than 80% of students will complete it

2 Put the following steps in the proper order for producing and administering a classroom test:

a. __________ Prepare the items
b. __________ Define the purpose of the test
c. __________ Choose the level of difficulty of the items
d. __________ Proofread
e. __________ Prepare instructions
f. __________ Order the items
g. __________ Set up an appropriate testing environment
h. __________ Assure sufficient accuracy
i. __________ Choose the types of item to use
j. __________ Choose the number of items to use
k. __________ Format the test

3 Based on your reading, label the following statements as either true (T) or false (F):

a. __________ Printing the instructions at the beginning of a test is appropriate for very young students.

b. __________ Double spacing should be used between items on a test.

c. __________ Teachers should not allow students to ask questions during a test.

d. __________ The easiest way to increase reliability is to have a small number of items on a test.

e. __________ Items with a moderate level of difficulty are ideal in achieving a large standard deviation.

4 Choose a chapter from a textbook in your area and describe a test that you would develop for a specific grade level. Consider all of the factors described in this chapter as you develop your test.

5 Follow-up question 4 above by actually preparing a test from either a list of objectives or from a Table of Specifications.

SPOTLIGHT ON THE CLASSROOM

Mrs. Anthony has been a 4th-grade teacher at Schaffer Elementary School in Leeland, Virginia, for the past 33 years. She understands the importance of developing quality tests for her students. The school purchased new science books for all of the grade levels for the upcoming school year. During the summer months, Mrs. Anthony has decided to read and prepare the new material. She wants to create all of the science tests that she will administer throughout the year.

The book is divided into chapters and units. Mrs. Anthony is developing a test that will be given at the end of each unit, or every three chapters. The tests will be summative evaluations that cover many objectives.

Mrs. Anthony feels that a variety of test items should be used on the tests. She plans to use matching questions for vocabulary, multiple-choice questions to test knowledge and comprehension, and one or two brief essays for slightly higher-level cognitive objectives. Since Mrs. Anthony teaches science to her homeroom class, time is not a factor. However, the children are only 10-years-old, so the test cannot be too lengthy, but still must measure a large number of objectives. She decides to put 10 matching questions, 15 multiple-choice questions, and two very brief essays on the tests. She believes that the tests should take the students about 30 minutes to complete. All of the items Mrs. Anthony plans to create are moderately difficult.

When it comes time to write the questions, Mrs. Anthony begins by referring to the tests in the teacher's edition sample questions. She is able to use only a few of the questions from the book because too many of them are too easy. She spaces the questions far apart, yet keeps the pages neat and even. At the beginning of each section, Mrs. Anthony places bold, clear directions. Finally,

Mrs. Anthony proofreads the tests she created and decides that they are ready to use for the new school year!

1. Discuss the test planning and development skills used by Mrs. Anthony.
2. Is there anything wrong or anything you would correct or add to her process?

NOTE

1. Separate scoring sheets are not recommended for younger students. They are simply too confusing.

HOW DO TEST FORMATS AND DEVELOPING TESTS APPLY TO MY PROFESSION?

Selecting the format for testing your students can have a tremendous impact on their success. Just as the type of test you use (described in Chapter 4) is important, the format for the test is also just as important. Regardless of which type and format of test, ensuring quality is the most important aspect. If the items violate the guidelines, then the results will not provide evidence for you to make appropriate conclusions for learning. Plan, plan, plan, and make sure that the tests align with the purpose of the assessment and what you are trying to determine with the results.

CHAPTER 6

Determining Quality of Assessments

The impacts of reliability and validity: Why are these so important?

AS YOU READ

As you read the chapter, think about the following:

- How do **you** define or explain test quality?
- What is **your** experience with test quality?

WHAT ARE THE GENERAL PRINCIPLES OF RELIABILITY AND VALIDITY?

Reliability

Reliability refers to repeatability or consistency or the ability to be replicated. A measure is considered to have evidence of reliability if it provides the same result over and over again. We cannot calculate reliability exactly—instead we can only estimate reliability. The instrument itself is not "reliable." The instrument (test) elicits scores that are then analyzed for evidence of reliability.

Internal reliability refers to the extent that data collection, analysis, and interpretations are consistent given the same conditions. For example, if there are two observers recording the same performance, do they agree when rating the same performance? If not, then internal reliability is lacking, the data become a function of who collects them rather than what actually occurred.

External reliability refers to the issue of whether or not independent research can replicate studies in similar settings given the same conditions. It is up to you to determine if the results are consistent if research is replicated.

Reliability Estimators

Reliability estimators are used to provide an approximation of the amount of random error around a true score. There are multiple reliability estimators that are used in measurement. Some of the estimators that are commonly used in educational settings as well as explanations about how to calculate the estimates and coefficients are provided in the bulleted list below.

The following are the most common reliability estimators and common approaches to calculating those estimates:

- **Inter-rater or Inter-observer reliability** is the degree to which different raters/observers give consistent estimates of the same phenomenon. There are multiple ways to calculate the reliability estimate when seeking results of inter-rater reliability. One common way is to score the level of agreement or disagreement between raters. This approach is something that is easily implemented in school settings. For example, as illustrated on the table below, there are two different raters judging three competitors in a science fair competition. All three judges were trained on a rubric that is used to guide their judgement scores for each of the contestants. The table below includes an example of calculating an inter-rater reliability estimate for three participants in a contest. Each participant was rated on a scale of one to five by two judges.

COMPETITOR	JUDGE 1 SCORE	JUDGE 2 SCORE	AGREEMENT
1	5	5	1
2	4	3	0
3	4	4	1

As evidenced on the table above, the judges agreed on the scores for both competitor one and competitor three. However, they did not agree on competitor number two. This is illustrated by using a one or zero on the agreement column of the table. To calculate the estimate of inter-rater reliability, you divide the total number of competitors (three in this case) by the number in agreement (two in this case). Then, you convert that number to a percentage.

Total number of competitors is 3
Total number in agreement is 2
Divide and convert to percentage
2/3 = 67% agreement

The researcher must determine the appropriate number of raters and the acceptable percent agreement.

The remaining reliability estimates often use various correlation coefficients to determine the relationship between the items on the instrument(s) based on the scores obtained from students' performance on the tests.

- **Test-retest reliability** assesses the consistency of a measure from one administration to another. Test-retest reliability estimates for two tests are often determined using the Pearson correlation coefficient. If you are seeking an estimate of test-retest reliability for more than two tests, use Intraclass Correlation.
- **Parallel-forms (equivalent forms) reliability** assesses the consistency of the results of two tests constructed in the same way from the same content area/domain. Generally, you seek the same mean and variance of obtained scores from the student results on each of the two tests. To obtain more sophisticated and potentially more accurate results, use a correlation coefficient to estimate reliability.
- **Split-half reliability** is a measure of reliability obtained by splitting items on a measure into two halves (in a random fashion) and obtaining a score from each half of the scale. The correlation between the two sets of scores is used as a measure of split-half reliability (Field, 2009). Similar to parallel-forms, if a general estimate of consistency will suffice, you can seek the same mean and variance to get an indication of whether or not the tests are similar.
- **Internal consistency reliability** assesses the consistency of results across items within a test. Specifically, we estimate the reliability of the results by determining the consistency of the results for multiple items designed to measure for the same construct. There are multiple ways to calculate estimates of internal consistency. These include: average inter-item correlation, average item total correlation, split-half reliability, and Cronbach's alpha (α). Cronbach's alpha is the average of all possible split-half correlations. Cronbach's alpha is the most frequently used estimator of internal consistency. Generally, teacher-created assessments should elicit a reliability estimate of approximately .70 or higher. Assessments with high-stakes associated should elicit higher internal consistency reliability coefficients of approximately .90 (professionally developed tests) or .80 for lower-stakes standardized tests (Wells & Wollack, 2003).

General comments regarding reliability estimators. There are multiple approaches used to calculate estimates of reliability. The following table of reliability estimators, including pros and cons of each, is an illustrative guide about reliability estimators.

RELIABILITY ESTIMATOR	PROS	CONS
Inter-rater	Best way to estimate reliability when your measure is an observation. There are various methods to calculating the level of agreement.	Multiple raters or observers are needed. Training the raters may be necessary and time consuming. The results are fairly rudimentary.
Test-retest	Single rater code the same measure on two different occasions. Feasible in experimental and quasi-experimental designs that use a control group. There are various methods to calculating the level of agreement but calculating correlation coefficients is the most common.	It provides no available information about reliability until posttest data are collected. If the reliability estimate is low, there is not much that can be done because typically the experiment is complete. Pearson correlation coefficient can overestimate relationships if there is a small sample size. Intraclass correlation is more challenging to calculate. Bias is a concern due to the knowledge of students about the content that will be included on the second administration of the test. The interval of time between administrations of the test impacts the estimate of reliability.
Parallel forms *(typically only used in situations where you intend to use both forms as alternate measures of the same thing)*	Two forms can be easily administered. Can avoid issues associated with re-testing using the same instrument.	It can be a challenge to have multiple items designed to measure the same construct. Analyses typically require the use of a correlational statistic that may pose challenges without appropriate resources. The assumption that the items are randomly placed onto each version of the test underpins this reliability estimate.
Split-half	Halving the test can be quickly accomplished.	It can be a challenge to have multiple items designed to measure the same construct. Analyses typically require the use of a correlational statistic that may pose challenges without appropriate resources.

RELIABILITY ESTIMATOR	PROS	CONS
Internal consistency	Most common method of examining reliability on tests that measure multiple constructs. This method only requires that one test be administered one time. Lower estimates (~.70) may be acceptable for teacher-created assessments versus standardized assessments (with high stakes).	It can be a challenge to have multiple items designed to measure the same construct. Analyses typically require the use of interpreting statistics that may pose challenges without appropriate resources.

Validity

Test responses provide **validity** evidence. The term validity refers to the truth of, correctness of, or degree of support for an inference (Cook & Beckman, 2006; Messick, 1980; 1994). When we say something is valid, we make a judgment about the extent to which relevant evidence supports that inference as being true or correct. We can never be certain that all of the many inferences drawn from a single experiment are true. In general, validity deals with the accuracy of the interpretations of the results (internal) and the generalizability of the results (external). Both types are matters of degree. It is difficult to have both high internal and external validity. Typically, if the causation (internal) is made, the generalizability (external) suffers. There are several types of validity and approaches to collecting evidence. The bulleted list below includes traditional perspectives on the various types of validity.

- **Face validity** is the extent to which the test appears to measure what it says it measures.
- **Content validity** the extent of how well the content of the test samples the class of situations or subject matter about which conclusions are to be drawn.
- **Criterion-related validity** is the extent to which test scores relate to one or more external variables (called criteria) which are considered to provide a direct measure of the characteristic or behavior in question. Generally, criterion-related validity has three categories: 1) predictive, 2) concurrent, and 3) construct.
    - **Predictive validity** indicates the extent to which individuals' future performance on the criterion (outcome) is predicted from prior test performance.

- **Concurrent validity** indicates the extent to which the test scores estimate individuals' present standing on the criterion.
- **Construct validity** is the degree to which certain theoretical or explanatory concepts or constructs account for performance on the test.

Modern Ideas of Validity

Measurement experts have different perspectives on validity and how many types there are. Construct validity seems to have the most controversy. As you see on the list above, traditionalists believe construct validity is a sub-set of criterion-related validity. However, modern ideas of validity argue that there is only construct validity and all other types are a sub-set of construct validity. Messick (1989) explains that validity is a unified whole that encompasses the ideas of test interpretation and use, based on evidence and consequence. Data from any source cannot be used independent from its context and theory (Messick, 1989; Moss 1992). Messick (1989) indicates that "validity always refers to the degree to which empirical evidence and theoretical rationales support the adequacy and appropriateness of interpretations and actions based on test scores. ... Inferences are hypotheses, and the validation of inferences is hypothesis testing" (p. 5).

According to Messick (1989), all validity evidence falls under the umbrella of **construct validity**. Evidence of construct validity should include the facets of test interpretation, value implications, relevance and utility of the test, decisions about the possible societal consequences, and implications for the use and interpretation of the test. Construct validity includes psychological and environmental factors in addition to the construct or content domain. Furthermore, validity evidence includes the inferences that are made from the data collected from various instruments. The intent of the test must be taken into context as well as one's personal experiences, testing conditions, and personality traits. Also, Messick (1989) argues that you cannot make inferences without considering the construct theory. He suggests that domain (content) and criterion can bolster test construction as well as test evaluation. Tests must be constructed in such a way as to be relevant for the purpose they are intended to serve, and representative of the skills/knowledge required. He adds that criterion measures must be evaluated in light of their construct validity. Messick (1989) argues that the appropriateness, meaningfulness, and usefulness of score-based inferences depend as well on the social consequences of testing. Therefore, the social values and social consequences cannot be ignored in considerations of validity.

Construct validity refers to the degree to which inferences are warranted from the observed persons, setting, and cause and effect operations sampled within a study to the constructs that these samples represent. It is fostered by a clear explication of the person, setting, treatment, and outcome constructs of interest; carefully selecting instances that match those constructs; assessing the match between instances and constructs; and revising constructs based on the study results. There are multiple threats to construct validity and these threats negatively impact any evidence of construct validity. The bulleted list below includes many threats to construct validity.

Regardless if one looks at validity from a traditional or modern perspective, there are multiple threats to construct validity. A list of those threats is provided below.

Threats to Construct Validity

The following list includes obstacles that may inhibit the ability to establish evidence of construct validity:

- **Underrepresentation:** The test does not include all facets necessary to measure a construct.
- **Construct-irrelevant test variance:** The test variance indicates that the test includes excess variance that is unnecessary for interpretation of the construct, such as construct-irrelevant difficulty (i.e., difficult language); construct-irrelevant easiness (i.e., too much time on a test).
- **Measurement variance:** Instrumentation changes can occur because of treatment and exposure to treatment may lead to a more complex design of the study. These differences are rarely examined by researchers across groups.
- **Confounding constructs with levels of constructs:** Sometimes an experimenter will draw a general conclusion about constructs when only some levels were actually studied.
- **Reactivity:** The factor structure and level of responses can be affected by whether a person is or is not accepted into the treatment group, even before that person receives treatment.
- **Novelty:** The introduction of an intervention can breed excitement, energy, and so on that can increase the possibility of success.
- **Compensatory rivalry:** The participants of the control group can feel competitive to do as well as the treatment group despite not getting treatment benefits.
- **Resentful demoralization:** Members of a group receiving a less desirable treatment, or no treatment can be resentful, which can change their responses to the outcome measures.

Ways to Test for Evidence of Validity

The content can be validated with expert judgments that support the extent to which the domain is covered on a test—not the extent to which inferences are made. Content validity actually does not qualify as validity at all, although such considerations of content relevance and representativeness clearly do and should influence the nature of score inferences supported by other evidence (Messick, 1989).

The construct can be validated by investigating what qualities a test measures by determining the degree to which certain explanatory constructs account for performance on the test. Classroom tests are not exact measures of constructs or content because often items that should be included are left off the measure altogether. When evaluating a test for evidence of validity, one should investigate the qualities of a test by determining the degree to which certain explanatory concepts or constructs account for performance on the test. Multiple measures of each construct measured on a test should be used to establish construct validation. Common methods for determining construct validity include item response theory (IRT) and reliability estimates. Additionally, exploratory factor analysis (EFA) and confirmatory factor analysis (CFA) are used to collect evidence of construct validity.

There are legitimate consequences to incorrectly interpreting construct scores (test results). Inferences are value-laden, which means that one's personal values can directly bias score-based inferences and actions, which can indirectly influence consequences for individuals, institutions, and society. When choosing a construct label, it is important to think about the values associated with various labels. There can be negative consequences to labeling constructs inappropriately. There can also be social consequences to various decisions related to test inferences. Messick (1989) suggests including some explanation for the critical value of contexts in which the facts are embedded.

Criterion-related validity is based on the degree of the empirical relationship, usually tested in terms of correlations or regressions, between test scores and outcomes. Relationships between the criterion and external variables are evaluated by comparing the test scores with one or more measurable external variables. For example, determining the degree to which test scores for one criterion is consistent with the scores of another measure or outcome (criterion) that has already proven evidence of validity. Simply stated, criterion-related validity is an estimate of the relationship between individuals' performance on two instruments purported to measure the same construct. Generally, testing for and collecting evidence of criterion-related validity is categorized as predictive or concurrent. **Predictive validity** includes the predictability of future scores on the criterion based on the results of present scores on the criterion. For example, the validity of an aptitude test for performance prior to college is the correlation between the aptitude test scores and college grade point averages (GPAs). The aptitude test scores are used to predict college GPA. **Concurrent validity** represents the extent to which the current test scores are related to an already validated measure. A simple explanation is validity that has evidence that it measures what it purports to measure based on the comparison to an instrument for which there is evidence that it measures what it is supposed to measure.

Convergent and Discriminant Evidence

When testing for evidence of validity, experts agree that both convergent and discriminant tests should be used (Campbell & Fiske, 1959). **Convergent evidence** signifies that the measure in question is coherently related to other measures of the same construct based on theoretical grounds. **Discriminant evidence** signifies that the measure is not related to exemplars of other distinct constructs that is critical for discounting other plausible hypotheses.

Internal Validity Versus External Validity

Validity is generally thought of in two ways. It is either internal or external. **Internal validity** refers to the ability to make causal inferences among the relationship between two variables (*X* and *Y*). Internal validity is not about reproducibility, inferences about the target population, or whether researchers measure what they think they measure. **External validity** refers to the generalizability of the inferences, that is, whether the causal relationship holds over variations in persons, settings, treatment variables, and measurement variables. In both cases, the researcher must be mindful of various threats to both internal and external validity. The threats are provided in the bulleted lists below.

Threats to Internal Validity

The following are the major threats to internal validity:

- **Ambiguous temporal precedence.** This concerns uncertainty about whether the cause precedes the effect. Sometimes cause can be bidirectional. For example, criminal behavior causes incarceration that causes criminal behavior that causes incarceration.
- **Selection.** This is the systematic differences between the average person in one condition relative to another. Random assignment can eliminate selection bias.
- **History.** This means that all events that occur between the beginning of the treatment and the posttest that could have produced the observed outcomes in the absence of that treatment. Possible remedies include participant isolation, selecting participants from the same general location, or ensuring the same schedule of testing.
- **Maturation.** Participants in the research project experience many natural changes that would occur even in the absence of treatment. Possible remedy includes ensuring groups are all roughly of the same age and are from the same location and culture.
- **Regression toward the mean.** A statistical tendency or phenomenon that occurs when there is an imperfect correlation. Multiple retakes of the same

exam will elicit different results from each administration. Extreme scores (high or low) will likely move closer to the mean as an individual retests. This tendency is in part a function of random measurement error and the strength of the correlation between variables. Possible remedy includes using measures with better psychometric properties to reduce error variance.

- **Attrition.** This is the failure of participants to complete all outcome measures. A special subset of selection bias in that those who drop out may influence the magnitude of observed differences between groups. Possible remedy includes matching post-treatment.
- **Testing.** This occurs when the initial test influences scores on a retest. Possible remedies include using IRT and Solomon Four group design where some receive a pretest and some do not.
- **Instrumentation.** This is possible temporal variance in an instrument used to measure the treatment effect.

Threats to External Validity

The following are the major threats to external validity:

- Interaction of the causal relationship with units.
- Interaction of the causal relationship over treatment variations.
- Interaction of the causal relationship with outcomes.
- Interaction of the causal relationship with settings.
- Context-dependent mediation.

Tradeoffs and Priorities With Internal Versus External Validity

Campbell and Stanley (1963) argued that "internal validity is the sine qua non," which means that internal validity is of utmost importance. Campbell and Stanley (1963) acknowledged that external validity (generalizability) is desired in educational research. However, internal validity and external validity are often at opposite ends of the spectrum. Thus, the higher the one, the lower the other. There are tradeoffs and priorities with internal and external validity. As the educator, one must make the decision about whether internal or external validity will take precedence over the other. These decisions are made consciously and should be considered when making consequential, educational conclusions about students.

WHAT DOES THE LITERATURE SAY ABOUT RELIABILITY AND VALIDITY?

Reading 6.1

TECHNICAL CRITERIA FOR EVALUATING TESTS

Vi-Nhuan Le and Stephen P. Klein

- How do we know if a test supports content standards?
- How do we know if test scores reflect students' actual learning?
- How can we measure the fairness of the tests and the reliability of test scores?

The previous chapter discussed features of test-based accountability systems. The focus of this chapter is on the technical quality of the tests themselves. In this chapter, we discuss the three main technical criteria by which the technical quality of tests should be evaluated: reliability, validity, and fairness. We begin by defining reliability. We then discuss certain factors that affect reliability, ways to quantify reliability, and the two most common reliability indices. Next, we describe judgmental and quantitative evidence that can be used to assess a test's validity. This is followed by a discussion of fairness, which includes issues of bias as well as of comparability in opportunities for students to learn and demonstrate their knowledge. We conclude by discussing the trade-offs among reliability, validity, fairness, testing time, and costs. (The specifics of how to calculate relevant psychometric statistics can be found in standard texts on this topic; e.g., Crocker and Algina, 1986.)

FIRST, SOME BACKGROUND

As background for what follows, it is important to appreciate that a test contains only a sample of all the questions that could be asked in a subject matter area or "domain," such as fourth-grade mathematics. Test scores are only useful to the extent that they accurately reflect the students' mastery of the domain the test is designed to measure. In other words, is it reasonable to infer from the test results how well the students would have performed if they had been asked a much larger and broader sample of questions from the domain? In short, how well can we generalize from the test (which is a sample) to the domain itself? And, are the scores reasonably free from bias and the effects of chance? The answers to these questions depend on several factors, including the way in which the test is designed and constructed and the ways in which the scores are used.

Because tests are samples, the confidence that can be placed in their scores depends on how well the test covers the range of knowledge and skills it is supposed to measure. No test can measure everything in a domain. However, a well-designed test can provide useful (albeit not necessarily comprehensive) information about the domain it is supposed to assess. Consequently, two tests that are designed to measure the same domain can differ substantially in content coverage, format, and problem types. And, these differences can affect the quality of their scores and the interpretation of what they mean.

Test quality hinges not only on how well the test is designed and constructed, but also on how its scores are used. For example, if scores are used to make high-stakes decisions about individual students (such as whether they receive a high school diploma), then it is imperative that these scores be stable indicators of student performance. Such consistency is not as critical when the students' scores are used to assess the outcomes of an educational program because the effects of chance are more likely to be balanced out across students within a program.

SCORE RELIABILITY

Reliability refers to the degree to which a test's scores are free from various types of chance effects. One way to think about this is to answer the question: What is the likelihood that a student's score, proficiency level, or pass/fail status would change if that student took the same test again the next day or took another version (i.e., another "form") of it?

Reliability is a characteristic of the scores on a test as distinct from a characteristic of its questions. For example, a test that is much too difficult for a particular group to answer (and results in everyone picking answers at random) will produce very unreliable scores for this group even though it may yield highly reliable scores for the students for which the test was intended. In this section, we discuss the factors that affect score reliability, the main quantitative indices of reliability, and how the decision about what constitutes "adequate" reliability depends on how the scores are used.

Factors That Affect Score Reliability

A student's score, proficiency level, or pass/fail status on a test can vary from one form of that test to another (or from one day to the next on the same form) for reasons that have nothing to do with any learning or maturation that took place in between the test sessions. The four broad classes of factors that can lower score reliability are item sampling, transitory variables, rater agreement, and test length and format.

Item Sampling

Various questions can be constructed to assess student knowledge of a given topic. Some of these questions can be put into one form of a test while others can be put into another form. Although both of these sets of questions are designed to be comparable, a given student may do better on one set than on another set while the reverse may be true for another student. In short, a student's score may fluctuate as a function of the particular version of the test taken. By definition, such inconsistency lowers reliability.

For example, suppose an examinee studied the contributions made to women's rights by Susan B. Anthony but not those by Elizabeth Cady Stanton. Suppose further that one form of the test covers the women's suffrage movement by asking questions about Ms. Anthony while another form has questions about Ms. Stanton. The student who studied Ms. Anthony's contributions is likely to do better on the first form than on the second one while the reverse is true for someone who learned only about Ms. Stanton's contributions. Thus, a student's score may depend on the particular sample of questions that are asked about a topic and the nature of a student's knowledge about that subject. This "student by task interaction" lowers score reliability because it is a matter of chance which form of the test a student takes.

Transitory Factors

There are a host of factors that have the potential for affecting student scores on a particular administration of a test that are independent of the particular sample of questions that are asked. These factors—all of which have the potential for lowering score reliability—include whether the student was healthy or ill, relaxed or anxious, or rested or tired on the day the test was given; the time of day the test was given; the room lighting and other environmental conditions during the test; the degree of adherence to instructions and time limits; the quality of the test booklets and other materials; the encouragement that teachers and others gave to students to do well on the test; and other such factors. The extent to which such transitory variables actually affect scores (and thereby lower reliability) depends on other factors, such as the nature of any disturbances that occurred during the test session and each student's response to such disturbances.

Rater Agreement

Raters (also called readers or judges) are needed to score student responses to essay questions, performance tasks, and other types of open-ended test questions. The reliability of a student's score on a test is reduced if these raters disagree with each other in the score they assign to an answer (or if a rater assigns different scores to the same answer on different occasions).

There are three basic ways in which raters can differ from each other. They can have different average scores (i.e., some raters are more lenient than others), score spreads (i.e., how much they use the full range of possible scores), or relative standings (i.e., the extent to which they agree on which answers are better than others). Corresponding differences can also arise when a rater evaluates the same answers on different occasions. Many studies have shown that a high degree of rater agreement can be obtained if there is a well-constructed and field-tested scoring guide (or "rubric") and if the raters receive extensive training in the use of these guidelines (Shavelson, Baxter, and Gao, 1993).

High rater agreement by itself does not mean that the students' test scores are highly reliable. For instance, readers can agree perfectly with each other on the scores that should be assigned to the answers the students wrote to essay question number 1, but the scores on this question may not correlate with the scores these students earned on the test's other essay questions.

Test Length and Format

All other factors being equal, longer tests (e.g., those with more questions) yield more reliable scores than shorter ones. And, obtaining a highly reliable estimate of an individual student's proficiency level often requires asking a large number of questions. The reason longer tests usually produce more reliable scores than shorter ones is that longer tests can contain a larger sample of questions from the domain being assessed. This is analogous to the benefits derived from conducting an exit poll at 20 precincts rather than just 5 precincts.

Tests that rely on essay questions or other types of open-ended tasks generally require far more testing time per student to achieve a given level of score reliability than do multiple-choice tests (Dunbar, Koretz, and Hoover, 1991; Linn, 1993; Wainer and Thissen, 1993). This is especially so when computer adaptive tests are used (Hamilton, Klein, and Lorie, 2000). However, multiple-choice tests cannot assess certain types of knowledge and skills. That is one of the reasons why score reliability is not the sole criterion for evaluating test quality.

Quantifying Score Reliability

For users of test data, arguably the most important reliability concern is the degree of confidence that can be placed on the scores. In other words, a key issue is how dependable are the scores? While there are a number of different methods used to estimate reliability, the most widespread procedures are test-retest, alternate forms, internal consistency, and generalizability theory. These methods provide a reliability statistic that quantifies the inconsistency of scores, namely a reliability coefficient or standard error of measurement (SEM). The following discussion is intended to be a conceptual introduction to ways in which to estimate and quantify reliability. Readers interested in a more technical discussion are referred to Feldt and Brennan (1989).

Methods of Estimating Reliability

There are three generic ways of estimating reliability: alternate forms, test-retest, and internal consistency. The *alternate forms method* involves having each student take two forms of the test, and correlating their scores on these forms. The *test-retest method* involves the students taking the same form of the test on two different occasions and examining the correlation between their scores on these occasions. The *internal consistency method* looks at the correlation among all the different questions within a single form of the test (such as between the even and odd numbered questions) or among the items within each subtest or section.

Each approach has its own set of assumptions, advantages, and limitations. Consequently, the reliability coefficient obtained with one method may differ substantially from the coefficient obtained with another method. Some of the factors that could lead to different coefficients are how much the scores depend on how fast the student answers, the degree to which the items within a test measure the same skills and knowledge, and the amount of time between test administrations.

A drawback of the reliability procedures just described is that they do not differentiate among multiple sources of error. For example, suppose two raters judged student performance on two different tasks that were administered on two different testing occasions. Using test-retest, alternate forms, or internal consistency methods, test users would not know the extent to which score inconsistencies were due to differences in raters judging the tasks, differences in the difficulty of the tasks, or differences in transitory factors (e.g., student fatigue) across testing occasions. This problem can be addressed by conducting a *generalizability* study that explores multiple sources of error simultaneously.

A typical generalizability study might involve students answering the same or comparable questions on different occasions, with two or more readers grading each student's answers to two or more questions. Such a study provides separate estimates for the various error sources (i.e., raters, tasks, occasions, and other sources of error). One of the strengths of a generalizability study is that it can identify the most serious sources of error, and this information can be used to guide decisions about what steps need to be undertaken in order to ensure more dependable scores. For instance, large error components associated with raters suggest that more reliable scores can be obtained by increasing the number of raters, the training of raters, or both. However, generalizability studies often involve a lot more grading and testing than is typically present in most assessment programs.

Reliability Statistics

After reliability is estimated, this information is reported via a reliability statistic. One of the most common reliability statistics is the *reliability coefficient,* which is an index that takes on any value from 0.0 to 1.0. A reliability coefficient of 0.0 indicates that the scores are completely undependable (i.e., the scores are no better than chance scores), whereas a coefficient of 1.0 indicates that the scores are completely dependable. For instance, a test that has perfect alternate-forms reliability means that an examinee's relative standing on one form of the test is entirely consistent with that examinee's standing on another form of that test. In practice, reliability coefficients are never 1.0, although all tests should aspire to yield coefficients as close to 1.0 as possible.

The second major reliability statistic is the *standard error of measurement* (SEM). The SEM is a function of the reliability coefficient and the degree to which the scores on a test tend to spread out around the average score (i.e., the standard deviation of these scores).[1] The SEM is analogous to a margin of error in a public opinion poll. For example, if a student's score on a test is 75 and the test has a SEM of 4, then the chances are roughly two out of three that the student's "true" score will fall between 71 and 79 (although 75 remains the best estimate of that score). The SEM is often useful for judging whether small differences in scores on a test are meaningful. For example, in this case, a score of 75 is fairly comparable to a score of 77 but is not comparable to a score of 80.

In general, the reliability coefficient is most useful for comparing the score reliabilities on different tests, whereas the SEM is especially useful for comparing scores within a test or for assessing the margin of error around a passing score (because the SEM is only meaningful in terms of the score scale on that test). As noted later in this section, what constitutes "adequate" reliability depends on several factors, including whether the scores are used to make decisions about individual students versus educational programs.

Score Reliability and Decisionmaking

As noted at the beginning of our discussion on reliability, one way to think about reliability is in terms of the likelihood that the test would result in the same pass/fail decision about a student (or put that student in the same proficiency category) regardless of the particular form of that test the student took (or whether the student took the same test at one time or another). A high degree of consistency in making that decision would coincide with the commonsense usage of the term reliability. However, the story is bit more complicated because the chance of making the same decision is also a function of the passing rate. Specifically, it is difficult to be inconsistent if almost everyone passes or almost everyone fails. On the other hand, score reliability is a particularly important consideration when the passing rate is anywhere between 25 and 75 percent.

Table 6.1 illustrates this relationship. It shows the percentage of students whose pass/fail status would change (i.e., from pass to fail or fail to pass) if they just took another form of the same test (or took the same test again). The table shows that the percentage of students whose pass/fail status would change is a function of both the passing rate and score reliability. For example, suppose a

TABLE 6.1 Percentage of Students Whose Pass/Fail Status Would Change as a Function of the Passing Rate and Score Reliability

PERCENT PASSING	SCORE RELIABILITY									
	0.00	0.10	0.20	0.30	0.40	0.50	0.60	0.70	0.80	0.90
90	19	18	17	16	15	13	12	11	9	6
80	32	30	28	27	25	23	20	17	14	10
70	42	39	37	35	31	29	25	22	17	12
60	48	45	42	39	36	32	29	25	20	14
50	50	47	44	40	37	33	30	26	21	14
40	48	45	42	39	36	32	29	25	20	14
30	42	40	38	35	32	29	26	22	18	13
20	32	30	28	27	25	23	20	18	14	10
10	19	18	17	16	15	13	12	11	9	6

Source: Klein and Orlando (2000).

large urban school district decides that it will promote third-grade students to the fourth grade only if they are able to earn a certain score on a reading test. Those who fail must go to summer school to take remedial instruction in reading. Suppose further that 70 percent of the district's 5,000 third graders meet this standard and that the test used to determine whether the standard is met has an alternate-forms reliability coefficient of 0.70 in this population of students.

Under the conditions just described, about 22 percent of the children would have their pass/fail status affected if they simply took another form of the same test. In other words, about 11 percent of the 5,000 third graders who were initially identified as having to participate in the summer school program would now be exempt from having to attend, whereas another 11 percent (or 550 children) would be in the opposite situation. Put another way, over a third of the 1,500 students who were required to take the summer program would not have been so classified if they had simply taken another form of the test. Increasing score reliability to 0.90 would cut the number of inconsistent classifications almost in half.

Examples like the one above illustrate the importance of score reliability when test scores are used to make important decisions about individual students. In contrast, score reliability is usually less a concern when scores are aggregated to the classroom or educational program level. This is because aggregating over large groups of examinees means that many (but not necessarily all) of the item sampling and transitory factors are balanced across students.

Whether test scores are considered "adequately" reliable depends on several factors, chief among them are the particular policies that are adopted for using those scores. Scores that are used to make decisions about individual students will require higher levels of reliability than scores that are used to make decisions about educational programs. For example, Koretz et al. (1994) found that the scoring of student portfolios in Vermont was too inconsistent to yield accurate interpretations about individual students, but that the results at the state level were reliable enough to monitor changes over time. Test users who wish to make decisions about individual students should be aware that a student's score is likely to vary much more across test administrations (i.e., have a much larger SEM) than will the mean of that student's classroom or school.

VALIDITY

Validity refers to the extent to which the scores on a test provide accurate information for the decisions that will be based on those scores (Cronbach, 1971; Messick, 1989). Validity is not an inherent characteristic of a test. It is a characteristic of the decisions that are based on the test scores and the subsequent uses of those scores (Cronbach, 1988). A given test may provide accurate information for one purpose but not for another. For instance, a test that requires examinees to multiply single-digit numbers can provide valid information about the multiplication skills of second graders, but it is not a valid measure of their proficiency in arithmetic because it does not assess addition, subtraction, or division skills. Moreover, if this same test is given to adults, it is probably more a measure of their perceptual speed than their multiplication skills.

The term validity is often used to encompass the reasonableness of both score-based inferences and the effects of a given use of a test, which are then referred to as "consequential validity." For clarity, we consider only the quality of inferences here. [...]

The kinds of interpretations that can be supported by test scores depend on the knowledge and skills that the test is supposed to assess. Thus, validity is necessarily related to test development. Test development consists of three main stages: *framework definition, test specification,* and *item selection* (National Research Council, 1998). In the framework definition stage, test developers use content standards to describe the scope of the domain to be assessed. The next stage involves the development of test specifications that delineate how the domain will be represented. In addition to circumscribing the format, time limit, and number of items that appear from each of the standards to be measured, test specifications may also describe the level and type of cognitive processes that students will need to draw upon in answering questions correctly. In the final stage, items are developed and selected to meet the test specifications. As will be discussed in the following sections, each of these stages affects the validity of the test.

The remainder of this section discusses the different types of judgmental and quantitative evidence that can be used to support the validity of a test's scores and the validity of scores and gains under high-stakes conditions.

Forms of Validity Evidence

In the past, it was common to think of the evaluation of tests as comprising several different "types" of validity tied to specific forms of evidence. For example, information on the appropriateness of test content was labeled "content validity" and information on the accuracy with which scores predict future performance was called "predictive validity." This terminology, however, has been largely abandoned because it is now generally recognized that the validity of an inference is a unified concept. Different types of evidence can contribute to that unitary judgment, and they should be taken together to evaluate the soundness of the interpretations with respect to the purposes of the test. There are still some instances in which people refer to specific types of validity; for example, the notion of "instructional validity," i.e., the degree to which students have had the opportunity to learn the content of an achievement test, still has wide currency.[2] However, these exceptions do not contradict the general notion that validity represents an integrative summary of evidence.

We next discuss different kinds of validity evidence, which are broadly categorized as those based on expert judgment and on quantitative data. This discussion is intended to be illustrative rather than comprehensive.

Evidence Based on Expert Judgment

A first step in evaluating the support a score provides for a given inference is to evaluate the content of the test. The framework and specifications for a test reflect the inferences the test designers intend the test to support. Thus, if the test is being used for purposes consistent with that intent, evaluation of the consistency of the test's content with its framework and specifications can provide useful validity evidence.

For state testing programs, the test specifications are designed to capture the state's content standards. Most state testing programs convene a group of content experts to ensure that the test specifications reflect the content standards, and that the test items reflect the test specifications. The steps are particularly important if commercially developed exams are used to monitor student progress toward state standards because such exams are not designed to reflect the content standards of any particular state. Some states, such as Delaware, use only test items that have been judged to address a content standard in their calculations of a standards-based score. Other states, such as California, augment the commercially developed exams with additional items that were developed to improve the overall alignment of the combined test with the content standards.

Because tests are samples from a content domain, they may omit or poorly represent some important aspects of the larger domain. This failure to capture the entire domain is called *construct underrepresentation*. If a test fails to capture important elements of the domain, scores can only justify narrow or qualified conclusions about performance. For instance, a test that is intended to assess knowledge of 19th-century American history but omits items about the Civil War provides an incomplete picture of students' knowledge of that period of history. Content experts can help safeguard against construct underrepresentation by identifying whether any of the excluded or deemphasized material represents important elements whose absence would seriously compromise the proposed interpretations.

Similarly, experts may find that the final set of test items includes content that is not pertinent to the intended inferences. This is called *construct-irrelevant content*. Some states, such as Maryland, assess student knowledge of mathematics with items that require written explanations. This may be a source of construct-irrelevant content because a low score in mathematics may indicate a lack of verbal skills to communicate an answer as opposed to a lack of mathematics knowledge (Hambleton et al., 2000). If that occurred it would undermine validity because a student who is actually comparable to a peer in terms of mastery of mathematics may receive a lower score solely because of differences in writing ability.[3] There also are issues regarding how much weight is given to different topics within a test (and because of statistical reasons, this weight may not coincide exactly with the number of items allocated to each topic).

Although content-related evidence is both essential and intuitively appealing, it has numerous limitations:

- First, test specifications are sometimes intentionally vague so as to avoid fostering overly narrow assessment goals (Popham, 1992).
- Second, the apparent content of an item, even when judged by an expert, is not always a trustworthy guide to the skills and knowledge that students will actually bring to bear in answering it. Research suggests it is difficult to determine the skills that are needed to answer a given item correctly by simply inspecting the surface features of the question (Hamilton, Nussbaum, and Snow, 1997). Instead, other methods such as interview procedures that ask students to "think aloud" as they solve the test items may be needed to provide evidence that the cognitive processes assumed to be underlying a given item are in fact elicited.
- Third, validity evidence about how well the test content corresponds with the test specifications may not be the most appropriate evidence in all situations. For example, when inferences attached to test scores are not those intended by the test's developers, the match between the test's framework and the test content is not as important as whether the test content can justify the *actual* (as opposed to intended) inferences.

  For example, the SAT I is designed to predict subsequent performance in college. Although some high school curriculum is included in the content, the SAT I is not intended to measure mastery of high-school curricula. Despite caveats from the sponsor of the SAT I that "it is completely inappropriate to use the [SAT I] as a measure of comparison between states, school systems, or schools" (Toch, 1984), the U.S. Department of Education nevertheless published the famous "wall charts," which used state average SAT I scores as a measure of the quality of states' educational systems (Ginsburg, Noell, and Plisko, 1988).[4] Because the content of the SAT I is not intended to represent what students are taught in school, scores cannot adequately support inferences about how well that material has been taught or learned (and there are large differences across states in the characteristics of the students who choose to take this test). Although the wall charts are gone, misuse of the SAT I and other admissions test scores continues to occur in various contexts. Thus, a key aspect of validation is clarifying whether the actual inferences drawn from test scores can be supported by the test content and its empirical relationships with other variables, as well as whether those inferences are consistent with the purposes for which the test was designed.

The above example underscores the importance of validating each proposed use of a test. Although it may be tempting to use an existing test for multiple purposes, a test developed and validated for one purpose cannot be assumed to be valid for another because the test content may not lend itself to interpretations other than the ones for which it was originally intended. Therefore, extreme caution should be exercised when test scores are used for purposes for which they have yet to be validated.[5]

Evidence Based on Quantitative Data

Because content evidence alone may not provide a sufficient gauge of the support a test score provides for a given inference, it is often important to examine empirical evidence based on actual performance on the test.[6] Two of the most important types of empirical evidence are (1) the pattern of relationships among the scores on different measures (including the one being investigated) and (2) the pattern of relationships among the items within a test.

Pattern of Relationships Among Measures. The kinds of skills and knowledge a test is thought to assess provide a guide to the patterns of relationships among measures that should be found. A test that purports to assess certain kinds of skills should show stronger relationships with other measures that assess those same skills than with measures that assess different skills (Campbell and Fiske, 1959). Scores from a mathematics test, for instance, should show stronger relationships with scores from another mathematics test than with scores from a reading test. This type of evidence, called *convergent-discriminant evidence,* has played an important role in recent years in evaluating large-scale assessments. Yen and Ferrara (1997) conducted a convergent-discriminant analysis of Maryland's state test, the Maryland School Performance Assessment Program (MSPAP), and found that MSPAP reading scores showed as strong a relationship to the math scores from the Comprehensive Test of Basic Skills/4 (CTBS/4) test as they had with the CTBS/4 reading scores. This result, which may have stemmed in part from the interdisciplinary nature of the MSPAP, raises possible concerns about the MSPAP's validity (Hambleton et al., 2000). Specifically, the math scores appear to be heavily influenced by a student's reading skills.

Depending on the inferences they are used to support, scores on tests can be compared with a variety of variables (i.e., not just the score on other tests). For example, the SAT I and the ACT composite are used to predict performance in college. Therefore, the most common approach to assessing the validity of inferences supported by these tests is to assess how well they correlate with grades in college. This is most often done by examining the prediction of freshman-year grade point average (GPA), but studies have also looked at cumulative GPA over a longer period (Baron and Norman, 1992), persistence in college (Wilson, 1980), and graduation rates (Astin, Tsui, and Avalos, 1996). It is also important to evaluate whether the criterion measure that the test is designed to predict (such as first-year grades in college) is itself a reliable indicator of student performance and whether the test provides equally accurate predictions across groups (because differential prediction may suggest test bias). This is discussed in more detail later in this chapter in the section on fairness.

Relationships Within Measures. Many testing programs report scores separately by skill or by knowledge type, often called "subtest" or "subscale" scores. It is not uncommon, for example, to find mathematics performance reported as two scores, such as "problem solving" and "computation." An implicit assumption in reporting sub-scale scores is that there are sets of items within the test that measure the distinct skills included in each subscale. In order to explore whether this assumption is warranted, an examination of the interrelationships among individual test items needs to be undertaken. If a test does indeed justify separate subscale scores, then items that measure the same skill will behave more like each other than like items on another scale. If all items show similar relationships to each other, then the test does not support interpretations based on sub-scale scores, and a total score should be reported instead.[7]

Validity of Scores and Gains Under High-Stakes Conditions

One area aspect of validity that has drawn increasing attention in recent years is the validity of inferences attached to high-stakes measures. A particularly serious threat to the validity of such inferences is the risk that the scores will become inflated. That is, increases in test scores are considerably larger than increases in student proficiency, so that inferences about student learning are biased. Score inflation can occur, for example, if teachers focus unduly on the specific content of a test at the cost of de-emphasizing untested material that is nevertheless important to the curriculum. Score inflation can also arise when teachers coach students by focusing on incidental aspects of a test, such as the particular rubrics used in scoring. Various forms of cheating also cause score inflation.

Score inflation is a particular concern when assessments are used to support inferences about *gains* in student performance. Under high-stakes conditions, scores may become progressively more inflated over time. Inflation of gains, however, can arise even in the absence of score inflation. For example, if an assessment is sufficiently novel, initial scores may be misleadingly low because students may not understand the directions or the kinds of responses that are expected of them. In this case, students' increasing familiarity with the test may mean that after one or two years scores are more accurate. Nonetheless, the *change* in scores from the starting point to the more accurate ending point would be exaggerated. In many instances, however, inflated gains arise from score inflation at the end of the period being considered, as opposed to unduly low scores when the assessment was first administered.

To evaluate possible score inflation, it is necessary to have some basis for comparing the level of scores. This can be done in at least two ways. In one district, Koretz et al. (1991) used national norms to place the results of two tests on the same scale and then compared mean scores across tests at one time. They also looked at trends over time within a single test. Both methods showed sizable score inflation. The more common approach, however, has been to compare trends on two tests. The logic of this approach is that gains on a high-stakes test should generalize to other measures. Although one would not expect perfect agreement (because even tests designed to measure similar domains differ), gains on a high-stakes test, if they are a valid indicator of improved student performance, ought to be reflected to a substantial degree in score gains on other measures that assess similar skills. For example, Koretz and Barron (1998) found that gains in mathematics scores in Kentucky during the 1990s on the high-stakes Kentucky Instructional Results Information System (KIRIS) assessment were nearly four times larger than the gains in Kentucky on the National Assessment of Educational Progress (NAEP). Similarly, Hambleton et al. (1995) found that Kentucky's initial rapid gains on the KIRIS reading assessment were not reflected in that state's NAEP reading scores. Klein et al. (2000) found similar results for the statewide achievement tests that are used in Texas.

Research on the inflation of gains remains too limited to indicate how prevalent the problem is. However, most studies that have specifically examined relevant evidence have found score inflation, and it has usually been severe. Thus, the research to date suggests that inferences based on scores from high-stakes tests should be validated using methods that are sensitive to score inflation.

FAIRNESS

Appropriate interpretations of test scores require that the test be fair to all takers. That is, the test should produce the same score for two test-takers who are of the same proficiency level. Unrelated characteristics of the test-takers, such as gender, ethnicity, or physical disabilities, and differences in administrative conditions should not affect the scores test-takers receive. Like validity, fairness is not an inherent property of the test but depends upon the use of the test. For a testing system to be fair, there are three key requirements: (1) the items and the test must be free of bias; (2) examinees must have comparable opportunities to demonstrate their knowledge and skills (i.e., there must be equitable treatment during the process of testing); and (3) examinees must have sufficient opportunity to learn the tested material (i.e., unless the purposes of the testing program include identifying students who have not had that opportunity). We next discuss each of these requirements.

Bias-Free Items and Tests

It is not uncommon for educators, policymakers, and the public to assume that a test is biased if there are large differences in average scores among groups. However, group differences in test scores may simply indicate that the groups differ in the knowledge, skills, and abilities that are being tested. Thus, freedom from bias does not necessarily entail score equality. By way of analogy, a fair test of some aspects of physical strength would not show that women and men are on average equally strong, but to be fair, it must give the same rating to individuals with comparable strength regardless of their gender.

Item Bias

Developers of most large-scale tests take a number of steps to detect and prevent bias on individual items. They have panels review the test to confirm that the items do not contain offensive or otherwise inappropriate language for various groups. They ensure that the exam corresponds to its test specifications and its curriculum frameworks or guidelines. They also may conduct statistical studies (i.e., *differential item functioning* [DIF] *analysis*) to flag questions that do not function in the same manner across groups. Specifically, DIF analysis flags an item when students who have the same total score on the test but belong to different groups have different probabilities of answering the item correctly. Although the presence of DIF in and of itself does not necessarily indicate that the item is flawed, some of the items flagged by the DIF analysis may be reviewed and excluded from the test before final scores are computed.

DIF analyses may be especially useful for suggesting sources of group differences on particular types of items. To illustrate, one of the items on a basic skills test for teachers in California asked, "How many half-pint servings are there in 5 gallons of milk?" White and black candidates performed better on this item than would be expected given their scores on the remaining items of the exam and the overall difficulty of this item among all candidates. The opposite was true for Hispanic and Asian candidates. It was hypothesized that Hispanic and Asian candidates were less familiar with the English system of measurement, and an item about the metric system would have shown the reverse pattern of group differences.

Whether information about group differences should be used in the test development process is open to debate. It is certainly true that the use of certain formats and content can affect the magnitude of group differences (Stocking et al., 1998; Willingham and Cole, 1997). For example, the extent to which gender differences in mathematics are manifested depends in part on the test's emphasis on reasoning skills (Harris and Carlton, 1993). Although some researchers believe it is the responsibility of test developers to select content that minimizes group differences while maintaining validity (Cole and Zieky, 2001), there has yet to be a consensus among the measurement community on whether group differences should be an explicit criterion when choosing test content.

Differential Prediction

A common method for examining bias in a test involves examining the predictive power of the test for different groups. This method is especially important when tests are used for the purpose of predicting who will do well in a future endeavor. For example, an increasing number of states are developing or are already using exams to certify that students have met the standards for high school graduation (Heubert and Hauser, 1999). There also is interest in using the scores on such tests for determining a student's eligibility for admission to state colleges and universities; i.e., as a replacement for the SAT I and ACT exams.

If high school graduation exams are to be used for these purposes, and if the large differences in scores (and passing rates) among racial/ethnic groups persist on these tests, then test developers will need to show that these differences do not stem from the exams being "biased" against the minority groups. One of the standard ways of examining this issue is to investigate whether minority and nonminority students with the same admission test scores do equally well in college. In general, if minority students do better, then the test is considered to be biased against them, whereas if they do less well than their nonminority classmates, then the test is biased in favor of the minority students. Previous research suggests that the latter outcome is much more likely to occur than the former (Linn, 1982).

Comparable Opportunities to Demonstrate Knowledge

Fairness requires that all examinees be afforded appropriate and equitable testing conditions. Standardization of administration conditions, which is common in most large-scale testing systems, is designed to ensure fair treatment by allowing all examinees to have the same amount of time to take the test, to receive the same instructions, and to take the test in comparable physical environments. In some instances, students are provided accommodations because they have disabilities or other challenges that prevent them from fully demonstrating their knowledge and skills under normal testing conditions. The accommodations are intended to offset the influence of construct-irrelevant factors on performance but, as we discuss next, have raised some validity questions in some instances.

Disabilities and Accommodations

The use of accommodations in large-scale testing has increased partly due to growing pressures to include all students in the assessment program and changes in the law stemming from the Americans with Disabilities Act. As is often mandated by a student's Individualized Education Plan (IEP), disabled students are granted accommodations to remove temporary or permanent disability-related barriers to performance that are ancillary to the construct the assessment is designed to measure. Test-based accountability policies have also contributed to increased attention to testing accommodations, as many states have sought to include students who had traditionally been exempted from testing requirements.

Currently, many state testing programs are struggling with ways in which disabled students can be accommodated. The major challenge is to balance students' needs for accommodations while maintaining comparability of score interpretations for students who have taken the test under standard conditions. A few states allow students to take state achievement tests under conditions that do not maintain score comparability (Consortium for Policy Research in Education, 2000). In Delaware, for example, some students are allowed to have the passages or text read to them during the reading test. Other states, such as Massachusetts and Maryland, have developed alternative assessments, such as portfolios, to be submitted in lieu of the state achievement test.

Relatively little is known about how testing accommodations affect score validity, and the few studies that have been conducted on the subject have had mixed results. Tippets and Michaels (1997) studied item interrelationships for accommodated and nonaccommodated students on the MSPAP and found the relationships to be comparable across the two groups. Similarly, Rock, Bennet, and Kaplan (1987) found that accommodated SAT I administrations were comparable to nonaccommodated administrations with respect to item interrelationships and reliability.

Other studies have found that accommodations may have been applied in ways that diminished validity. In one such case, disabled subgroups who had received accommodations demonstrated implausibly high scores, outperforming nondisabled students by nearly one-half a standard deviation (Koretz, 1997). The effect disappeared when the test was administered two years later, but it is impossible to determine what caused the change (Koretz and Hamilton, 2000). More research is needed to understand the sources of these findings.

Limited English Proficient Students and Accommodations

Because every test necessarily entails language to some degree, the performance of limited English proficient (LEP) students reflects factors that are ancillary to the intended construct and may result in underestimating their knowledge.[8] To circumvent this problem, many state testing programs offer accommodations that range from extending time limits to allowing the test instructions to be read orally in the student's dominant language (Consortium for Policy Research in Education, 2000).

A few states, including Vermont, New York, and New Mexico, offer versions of their assessments in languages other than English. Administering non-English versions of a measure has proved problematic, particularly because of equivalence problems. A translated test may not be comparable in content or rigor (and hence validity) as the original version because many words do not have the same frequency rates or difficulty levels across languages. Studying the comparability of an NAEP test that had been translated to Spanish, Anderson, Jenkins, and Miller (1996) concluded that comparisons across different language versions of a test were not possible because of substantial translation problems, including confusing language, varying vocabulary meanings, and increased potential for misunderstanding by LEP students. Other studies have reported that translated versions may be longer than the English version, thereby placing more demands on memory (Valencia and Rankin, 1985).

Beyond lack of translation equivalency, it cannot be assumed that a score obtained from the translated version should be interpreted in the same way as if it were obtained from the original test because test items may not function in the same manner. Studies have shown that psychometric properties are not always robust across the two languages, and in some cases, the translated items show poorer technical quality (Anderson, Jenkins, and Miller, 1996). As with the research on students with disabilities, more work is needed to understand how to interpret scores from these tests and how to develop tests that provide valid measures of achievement for all students.

Opportunities to Learn

For tests that are used to deny or award students a high school diploma or that have other high-stakes outcomes for students, equity concerns require that all examinees have had sufficient exposure to the material on the test. The landmark Debra P. v. Turlington (1981) decision established that legally defensible high school graduation tests must have "instructional or curricular validity."[9] That is, students should have had an opportunity to learn the skills and gain the knowledge that the test is designed to measure.

It is unclear how to evaluate whether students have had an opportunity to learn the tested material. Some believe instructional validity should hinge on the actual instruction students receive rather than the formal written curriculum (Airasian and Madaus, 1983). In other words, it is not enough for the test content to reflect the content standards; instead, there must be evidence that the content was actually delivered within the classrooms. However, given that it may be impossible to show that every student was exposed to the necessary content, it has been argued that aligning tests to the published content standards is sufficient for demonstrating opportunity to learn. These issues are far from resolved, and the increasing emphasis on student-level accountability may lead to additional litigation.

Opportunity-to-learn issues are relevant not only to individual students but also to schools and school systems. Differences in test scores across schools with students who seem to be similarly situated may signal differences in educational opportunity. For that and other similar reasons, a state may want to administer tests to all students in order to help identify educational programs that are not providing sufficient educational opportunities. The fact that some students have not received opportunities to learn the content covered on a test should not prevent the testing of that content, although this lack of opportunity does influence the interpretation and use of the test scores. In this situation, scores may provide useful information about the effectiveness of different school programs, but in the interest of fairness, the scores should not be used to make high-stakes decisions about individual student mastery.

TRADE-OFFS: TESTING TIME, COSTS, AND OTHER FACTORS

How test scores are used affects decisions about test design, which in turn affect decisions about content coverage, test length, format (multiple-choice versus essay and other open-response formats), administration and scoring procedures, and other factors. And, all of these decisions are likely to affect reliability, validity, fairness, testing time, costs, and other variables. For example, if scores are used to make high-stakes decisions about individual students, such as whether they are promoted, then a high degree of score reliability is needed. However, obtaining the score reliability that is needed to make decisions about individuals usually requires a substantial amount of testing time per student (which drives up costs) and/or narrowing the range of content standards that are assessed (such as not assessing those standards that require open-ended responses).

In contrast, if the scores are used to assess program effectiveness, then adequate score reliability can often be obtained with far less testing time per student and with greater validity. This result is achieved by having some students answer one set of questions and having other students answer other sets. [...] This strategy, which is called "matrix sampling," may increase validity because it facilitates obtaining a more comprehensive assessment of the students' knowledge and skills with less testing time than would be feasible had every student responded to every question. That is why a number of large-scale testing programs, such as NAEP, use matrix sampling. It is also why these programs do not report scores for individual students; i.e., such scores are not reliable enough for making decisions about individuals because a given student may see questions from only a small proportion of the content domain that the total test is assessing.

When test scores are used for making decisions about both individuals and programs, and when the testing time and other resources that are available for assessment activities are limited (as they almost always are), then there has to be some compromise among score reliability, validity, and costs. For example, a district may purchase a commercially developed test that is not especially well aligned with its standards, but which provides reliable scores in a relatively short amount of testing time and at a price the district can afford.

Tasks such as essays or performance assessments may be seen as more "authentic" indicators of student performance than multiple-choice tests and they can assess certain skills that are not measured well with the multiple-choice format. However, compared with multiple-choice items, the answers to open-response questions are much more costly to score (Stecher and Klein, 1997) and those scores are generally less reliable per hour of testing time (Wainer and Thissen, 1993).

Trade-offs in test design may also affect fairness. Girls, for example, tend to earn higher scores than boys on essay tests while the reverse is true on multiple-choice tests (Bolger and Kellaghan, 1990; Breland et al., 1994; Mazzeo, Schmitt, and Bleistein, 1993). Decisions about test design can therefore influence differences in scores among groups (Klein et al., 1997).

Issues regarding trade-offs with fairness also arise in decisions about releasing test questions to the public. For instance, many policymakers would like to have parents and teachers review a test before it is used so that they can provide input on what is assessed. Others would like a test's questions disclosed at least after the exam is administered so that they can check the accuracy of scoring keys and rubrics as well as to have guidance for better aligning curriculum and instruction with what is assessed. Indeed, that was how a significant error was discovered in the scoring key on a state's high school graduation test. However, releasing questions can lead to abuses that seriously undermine test validity (Cizek, 1998). For example, if in the weeks leading up to a state exam, students are taught how to solve the particular types of math problems that appear on this test, then their scores on the exam will reflect this specialized knowledge, but contrary to the purpose of the test, these scores will not generalize to the larger content domain that the test is designed to measure. In short, the scores will not be valid indicators of the students' mastery of the domain.

One partial solution to the disclosure dilemma that strives to maintain both validity and fairness involves (1) releasing a sample of the test's questions, (2) adding further safeguards to protect against scoring key problems, and (3) calibrating the next version of the test to the previous one through a set of secure items that are used on both tests. Another approach is to use a computer-based testing system in which students in the same classroom answer different questions that are drawn from the same bank of thousands of questions that are released but constantly updated and expanded (Hamilton, Klein, and Lorie, 2000).

CONCLUSION

With the increasing reliance on test scores as instruments of educational policy reform, it becomes more important than ever that tests are technically sound. To conclude this chapter, we summarize some of the key points in evaluating and using tests.

Tests should be evaluated relative to the purpose for which the scores are used. A test that is valid for one purpose may not be valid for another purpose. For example, tests that are used to evaluate the effectiveness of a given educational program may not be appropriate for making high-stakes decisions about individual student mastery. Similarly, test scores that are considered reliable enough to hold schools accountable may not be sufficiently reliable to make decisions about individual students.

Test scores are not a definitive measure of student knowledge or skills. An examinee's score can be expected to vary across different versions of a test because of differences in the particular sample of items that are asked, differences in the way graders evaluate student responses, and differences in transitory factors, such as the examinee's attentiveness on the day the test was taken. For these reasons, no single test score can be a perfectly dependable indicator of student performance, and high-stakes decisions about individuals should be based on factors other than the score on a single test (American Educational Research Association, American Psychological Association, and National Council on Measurement in Education, 1999).

It is important to recognize that decisions about test design (i.e., decisions on test length, item format, content coverage, and other considerations) require trade-offs with respect to reliability, validity, fairness, and costs. For instance, tests that include essay items may assess certain skills that cannot be measured with multiple-choice items, but are also more costly and less reliable per hour of testing, and in some cases may give female test-takers a relative advantage over male test-takers. Longer tests are generally more reliable and can sample a wider range of content than shorter tests. They also can provide more diagnostic information about student strengths and weaknesses. However, longer tests require more student testing time, which is time that is typically taken away from instruction. Thus, decisions about test design should be informed by the ways in which scores will be used and by resource constraints as well.

The issues discussed in this chapter regarding reliability, validity, and fairness highlight some of the many challenges that are faced by those who develop measures, implement test-based accountability systems, and use test results to make decisions about students and educational programs. Because test scores are numeric, they are often assumed to have a degree of precision that is not always justified. That is why those who use test scores to make decisions need to evaluate the technical characteristics of those scores relative to the purpose or purposes for which the scores are used. This evaluation is needed to make informed decisions about what the scores actually mean and how much confidence can be placed in them.

NOTES

1 The size of the SEM for a given score depends on how close that score is to the test's mean score. The further the score is from the mean, the larger its SEM.

2 Some measurement experts argue that "instructional validity," while important, should not be considered an aspect of validity at all; whether a test adequately supports an inference about students' proficiency in a given

area does not depend on whether the students were taught that material in school. In practice, whether instructional validity is relevant is in large part a function of the kinds of inferences that are being made on the basis of scores (e.g., whether scores are interpreted as indicators of student academic attainment or of school effectiveness).

3 Whether assessing mathematics achievement with written explanations constitutes construct irrelevance depends upon how mathematics achievement is defined. If mathematics achievement is defined to include the ability to communicate mathematics concepts in writing, then writing would probably *not* be considered construct-irrelevant content.

4 During this time, the SAT I was known as the SAT.

5 In some instances, it may be necessary to conduct reliability and validity studies after the test has been administered, but before the scores are used to make decisions about individuals or programs.

6 Although there are published standards to guide validity investigations, there are no clear-cut rules regarding what evidence is important in what contexts. For example, in a court case related to the California Basic Educational Skills Test for teachers, the federal court held that content validation alone was sufficient for establishing the validity of a certification or employment test; i.e., criterion-related evidence was not required to determine whether the test scores were valid for the purposes for which they were being used (Mehrens, 1999).

7 There are exceptions to this principle. For example, while the pattern of correlations among items may coincide with the intended subscales when the student is the unit of analysis, it still may be possible that the subscales may vary as a function of schools or educational programs.

8 Language and reading tests are exceptions to this concern.

9 See Phillips (1993) for a discussion of the *Debra P. v. Turlington* court decision.

REFERENCES

Airasian, P. W., and G. F. Madaus. (1983). Linking testing and instruction: Policy issues. Journal of Educational Measurement, 20 (2), 103–118.

American Educational Research Association, American Psychological Association, and National Council on Measurement in Education. (1999). Standards for Educational and Psychological Testing. Washington, D.C.

Anderson, N. E., F. F. Jenkins, and K. E. Miller. (1996). NAEP Inclusion Criteria and Testing Accommodations. Findings from the NAEP 1995 Field Test in Mathematics. Washington, D.C.: Educational Testing Service.

Astin, A., L. Tsui, and J. Avalos. (1996). Degree Attainment at American Colleges and Universities: Effect of Race, Gender, and Institutional Type. Washington, D.C: American Council on Education.

Baron, J., and M. F. Norman. (1992). SATs, achievement tests, and high-school class rank as predictors of college performance. Educational and Psychological Measurement, 52, 1047–1055.

Bolger, N., and T. Kellaghan. (1990). Method of measurement and gender differences in scholastic achievement. Journal of Educational Measurement, 27, 165–174.

Breland, H. M., D. O. Danow, H. D. Kahn, M. Y. Kubota, and M. W. Bonner. (1994). Performance versus objective testing and gender: An exploratory study of an Advanced Placement history examination. Journal of Educational Measurement, 31, 275–293.

Campbell, D. T., and D. W. Fiske. (1959). Convergent and discriminant validation by the multitrait-multimethod matrix. Psychological Bulletin, 56, 81–105.

Cizek, G. J. (1998). Filling in the blanks: Putting standardized tests to the test. Fordham Report, 2(11). http://www.edexcellence.net/library/cizek.pdf (last accessed May 15, 2002).

Cole, N. S., and M. J. Zieky. (2001). The new faces of fairness. Journal of Educational Measurement, 38(4), 369–382.

Consortium for Policy Research in Education. (2000). Assessment and Accountability Systems: 50 State Profiles. http://www.cpre.org/Publications/Publications_Accountability.htm (last accessed May 13, 2002).

Crocker, L., and J. Algina. (1986). Introduction to Classical and Modern Test Theory. Fort Worth, Tex.: Harcourt Brace Jovanovich.

Cronbach, L. J. (1971). Test validation. In R. L. Thorndike (ed.). Educational Measurement, 2nd ed., 443–507. Washington, D.C.: American Council on Education.

Cronbach, L. J. (1988). Five perspectives on validity argument. In H. Wainer and H. I. Braun (eds.). Test Validity, 3–17. Hillsdale, N.J.: Erlbaum.

Debra P. v. Turlington, 474 F. Supp. 244 (M.D. Fla 1979); aff'd in part, rev'd in part, 644 F.2d 397 (5th Cir., 1981).

Dunbar, S. B., D. Koretz, and H. D. Hoover. (1991). Quality control in the development and use of performance assessments. Applied Measurement in Education, 4(4), 289–304.

Feldt, L. S., and R. L. Brennan. (1989). Reliability. In R. L. Linn (ed.), Educational Measurement, 3rd ed., 105–146. New York: Macmillan.

Ginsburg, A. L., J. Noell, and V. W. Plisko. (1988). Lessons from the Wall Chart. Educational Evaluation and Policy Analysis, 10(1), 1–12.

Hambleton, R. K., J. Impara, W. Mehrens, and B. S. Plake. (2000). Psychometric Review of the Maryland School Performance Assessment Program (MSPAP). Maryland State Department of Education. Baltimore. http://www.abell.org/ (search "Publications" link; last accessed May 13, 2002).

Hambleton, R. K., R. M. Jaeger, D. Koretz, R. L. Linn, J. Millman, and S. E. Phillips. (1995). Review of the Measurement Quality of the Kentucky Instructional Results Information Systems, 1991–1994. A report prepared for the Office of Education Accountability, Kentucky General Assembly. Frankfort, Ky.: Office of Education Accountability.

Hamilton, L. S., S. P. Klein, and W. Lorie. (2000). Using Web-Based Testing for Large-Scale Assessments. IP-196. Santa Monica, Calif.: RAND.

Hamilton, L. S., E. M. Nussbaum, and R. E. Snow. (1997). Interview procedures for validating science assessments. Applied Measurement in Education, 10(2), 181–200.

Harris, A. M., and S. T. Carlton. (1993). Patterns of gender differences in mathematics items on the Scholastic Aptitude Test. Applied Measurement in Education, 6, 137–151.

Heubert, J. P., and R. M. Hauser (eds.). (1999). High Stakes: Testing for Tracking, Promotion, and Graduation. Washington, D.C.: National Academy Press.

Klein, S. P., L. S. Hamilton, D. F. McCaffrey, and B. M. Stecher. (2000). What Do Test Scores in Texas Tell Us? IP-202. Santa Monica, Calif.: RAND.

Klein, S. P., J. Jovanovic, B. M. Stecher, D. McCaffrey, R. J. Shavelson, E. Haertel, G. Solano-Flores, and K. Comfort. (1997). Gender and racial/ethnic differences on performance assessments in science. Educational Evaluation and Policy Analysis, 19, 83–97.

Klein, S. P., and M. Orlando. (2000). CUNY's Testing Program: Characteristics, Results, and Implications for Policy and Research. MR-1249-CAE. Santa Monica, Calif.: RAND.

Koretz, D. (1997). The Assessment of Students with Disabilities in Kentucky. CSE Technical Report 431. Los Angeles: National Center for Research on Evaluation, Standards, and Student Testing.

Koretz, D., and S. I. Barron. (1998). The Validity of Gains on the Kentucky Instructional Results Information System (KIRIS). MR-1014-EDU. Santa Monica, Calif.: RAND.

Koretz, D., and L. S. Hamilton. (2000). Assessment of students with disabilities in Kentucky: Inclusion, student performance, and validity. Educational Evaluation and Policy Analysis, 22,255–272.

Koretz, D., R. L. Linn, S. B. Dunbar, and L. A Shepard. (1991). The effects of high-stakes testing on achievement: Preliminary findings about generalization across tests. Paper presented at the annual meeting of the American Educational Research Association,Chicago.

Koretz, D., B. Stecher, S. Klein, and D. McCaffrey. (1994). The Vermont portfolio assessment program: Findings and implications. Educational Measurement: Issues and Practice, 13(3), 5–16.

Linn, R. L. (1982). Ability testing: Individual differences, prediction, and differential prediction. In A. K. Wigdor and W. R. Garner (eds.). Ability Testing: Uses, Consequences, and Controversies, 335–388. Washington, D.C.: National Academy Press.

Linn, R. L. (1993). Educational assessment: Expanded expectations and challenges. Educational Evaluation and Policy Analysis, 15(1), 1–16.

Mazzeo, J., A. P. Schmitt, and C. A. Bleistein. (1993). Sex-Related Performance Differences on Constructed-Response and Multiple-Choice Sections of Advanced Placement Examinations. CB Report No. 92-7; ETS RR-93-5. New York: College Entrance Examination Board.

Mehrens, W. A. (1999). The CBEST saga: Implications for licensure and employment testing. The Bar Examiner, 68, 23–31.

Messick, S. (1989). Validity. In R. L. Linn (ed.). Educational Measurement, 3rd ed., 13–103. New York: Macmillan.

National Research Council. (1998). Uncommon Measures: Equivalence and Linkage Among Educational Tests. M. J. Feuer, P. W. Holland, B. F. Green, M. W. Bertenthal, and F. C. Hemphill (eds.). Committee on Equivalency and Linkage of Educational Tests. Washington, D.C.: National Academy Press.

Phillips, S. E. (1993). Legal Implications of High-Stakes Assessment: What States Should Know.Michigan State University, North Central Regional Educational Laboratory, Illinois.

Popham, W. J. (1992). A tale of two test-specification strategies. Educational Measurement: Issues and Practice, 11(2), 16–17, 22.

Rock, D. A., R. E. Bennett, and B. A. Kaplan. (1987). Internal construct validity of a college admissions test across handicapped and non-handicapped groups. Educational and Psychological Measurement, 47, 193–205.

Shavelson, R. J., G. P. Baxter, and X. Gao. (1993). Sampling variability of performance assessments. Journal of Educational Measurement, 30, 215–232.

Stecher, B. M., and S. P. Klein. (1997). The cost of science performance assessments in large-scale testing programs. Educational Evaluation and Policy Analysis,19, 1–14.

Stocking, M., T. Jirele, C. Lewis, and L. Swanson. (1998). Moderating possible irrelevant multiple mean score differences on a test of mathematical reasoning. Journal of Educational Measurement, 35, 1–24.

Tippets, E., and H. Michaels. (1997). Factor structure invariance of accommodated and non-accommodated performance assessments. Paper presented at Annual Meeting of the National Council on Measurement in Education. Chicago.

Toch, T. (January 11, 1984). E.D. issues study ranking states on education. Education Week. http://www.edweek.org/ew/ewstory.cfm?slug=05280012.h03&keywords=Toch (last accessed May 13, 2002).

Valencia, R. R., and R. J. Rankin. (1985). Evidence of context bias on the McCarthy Scales with Mexican-American children: Implications for test translation and nonbiased assessment. Journal of Educational Psychology, 77(2), 197–207.

Wainer, H., and D. Thissen. (1993). Combining multiple-choice and constructed response test scores: Toward a Marxist theory of test construction. Applied Measurement in Education, 6, 103–118.

Willingham, W. W., and N. S. Cole. (1997). Gender and Fair Assessment. Mahwah, N.J.: Erlbaum.

Wilson, K. (1980). The performance of minority students beyond the freshman year: Testing the "late bloomer" hypothesis in one state university setting. Research in Higher Education, 13(1), 23–47.

Yen, W. M., and S. Ferrara. (1997). The Maryland School Performance Assessment Program: Performance assessment with psychometric quality suitable for high stakes usage. Educational and Psychological Measurement, 57(1), 60–84.

HOW DO RELIABILITY AND VALIDITY APPLY TO MY PROFESSION?

This chapter provided an overview of the types of reliability and validity, as well as some basic principles for collecting evidence of each. Estimates of reliability and validity are two of the most important concepts related to testing and measurement in education. As an educator, it is your responsibility to do no harm. As such, making decisions based on the results of assessments (whether teacher created or commercially developed) happens on a daily basis. It is imperative that any decision that is made is based on assessments that elicit evidence of both reliability and validity. It would be irresponsible if you were to make decisions about students based on tests that you haphazardly created to assess student knowledge of the content you believe you thoroughly taught. Thus, as educators, it is our responsibility to ensure that any consequential decisions that we make about or for our students are made using instruments for which we have evidence of both reliability and validity.

CHAPTER 7

Item Analysis

So, you think you created good items, but how do you know?

AS YOU READ

As you read the chapter, think about the following:

- How do **you** define or explain item analysis?
- What is **your** experience with item analysis?

WHAT ARE THE GENERAL PRINCIPLES OF ITEM ANALYSIS?

Item Analysis Statistics

There are multiple statistics that can be applied to determine the quality of the items on a test. When there are dichotomous responses (correct vs. incorrect), there are two very common ways to evaluate the quality of the item or question itself. The most common item analysis statistics include the difficulty index and the discrimination index. The general purpose of item analysis is to determine the level of difficulty for each item and the item's ability to discriminate between high scorers and low scorers on a particular test. Generally, items of moderate difficulty have the potential for good item discrimination.

Item analysis is a post hoc analysis performed to compare the performance on each item of the most and least successful examinees on the total test. This analysis will allow you to identify items that are nonfunctional, intrinsically ambiguous, or miskeyed.

Item Difficulty Index

The item difficulty index (p) is actually the proportion of students (test-takers) who answered a particular test question correctly. It is often reported as a *p-value*. *P-values* range from 0.0 and 1.0. The lower the value of the proportion value (difficulty index), the fewer the number of correct responses. The higher the value, the more correct responses. It is important not to conclude that the item is too difficult or too easy. Instead, reflect on how well (thoroughness) you covered the material. Also, look at the wording of the test question. Perhaps the question was confusing. Thus, few students answered correctly. Typically, criterion-referenced tests will have higher *p*-values (~0.9) than norm-referenced tests (*p*-value ~0.5) because norm-referenced tests are designed to be more difficult and to spread out the scores of the students or test-takers.

To determine the difficulty of a particular test item, complete the following steps:

- **Step 1:** Analyze the results for each item and categorize them into two groups. Use a label of "1" for those that answered correctly and "0" for those that answered incorrectly (regardless of the number of answer choices for each item).
- **Step 2:** Divide the number of students who pass the particular item that you are analyzing by the total number of test-takers in the group (the number that answered that particular item/question). This will give you the item difficulty index.
- **Step 3:** Repeat Step 1 until you've analyzed each item that you're interested in analyzing.

The following table is an example for 10 students that completed a three-item test.

STUDENT ID	ITEM 1	ITEM 2	ITEM 3
A	1	0	1
B	0	1	0
C	0	1	0
D	1	1	0
E	1	1	0
F	1	1	0
G	0	1	1
H	0	0	1
I	1	0	1
J	1	0	0
Number correct	**6**	**6**	**4**
Total number that answered question	**10**	**10**	**10**
Difficulty Index (*p*)	**0.60**	**0.60**	**.040**

General Guidelines for Acceptable Difficulty Indices

Generally agreed-upon guidelines for analyzing item difficulty include the following:

- $p > 0.90$: Omit or revise the item; If the question elicits a correct response from more than 90% of students, it is likely a concept that is not worth testing.
- $p < 0.20$: Omit or revise the item; The item language should be reviewed to ensure the question is worded appropriately. If you believe the question is appropriate and well written, think about modifying your instruction when teaching this concept.

Guessing. To determine the probability of guessing the correct answer, simply divide one by the number of possible answer choices. For example, if you have a multiple-choice question with four possible responses, then there is a one in four chance (25%) to guess the correct answer. If the question is true/false, then there is a 1 out of 2 (50%) chance of guessing the correct answer. The letter g represents the chance level meaning the item was correctly answered simply by chance (guessing).

Calculating Item Difficulty Optimal Values. When analyzing multiple choice items, the number of possible responses impacts the ideal level of difficulty for purposes of discriminating between high and low achievers. There is a fairly straightforward mathematical way to calculate an optimal level for acceptable *p-values*. To compute this level, first look at the number of possible answer choices for a particular question. We will use an example with four answer choices.

Knowing that the item difficulty ranges from 0.0 to 1.0 and that if everyone answers correctly, the *p*-value would be 1.0. If nobody answers correctly then the *p*-value would be 0.0.

In the case above, there are four options. That means that simply by chance, there is a 25% likelihood that the student will answer the item correctly regardless of his knowledge ($g = 0.25$). To calculate the optimal difficulty value for that particular item, add the probability that everyone answered correctly 1.0 to the level of "chance" $g = 0.25$ (total = 1.25). Then divide that by 2 ($1.25/2 = \sim 0.625$). Thus, in this case, the optimum *p*-value is 0.625 which is the same as averaging the 1.0 and the .25. As the number of answer choices increases, the optimal *p*-value decreases.

The following table shows the generally accepted guidelines for optimal difficulty based on the number of possible responses.

NUMBER OF ANSWER CHOICES	OPTIMAL DIFFICULTY
Two	0.75
Three	0.66
Four	0.62
Five	0.60

Item Discrimination. Item discrimination index (*d*) is used to distinguish between test-takers who know the content and those who don't or high achievers versus low achievers. The item discrimination index is actually a correlation (strength of relationship) estimate—specifically, a point biserial correlation coefficient. It can range from –1.0 to 1.0. A strong and positive correlation indicates that students who get any one particular question correct will generally have a high score on the overall exam. If you think about this, it makes sense. If you randomly select the results of one particular question on a test to analyze and a particular student gets that particular question correct; then there is a higher probability that he will do well on the entire test. Think about it this way: If a student knows the content and performs well on the entire test, then that student is likely to know the overall content that is being tested. If students performed well on a test and they did *not* know the content well then there is likely something wrong with the test or the test questions. The higher the discrimination index, the higher the percentage of the students who scored high on the test correctly answered the item under study.

When *d* is 1.00, it means that *all* students in the upper group (top 27% or top 50%) and zero or no students in the lower group (bottom 27% or bottom 50%) correctly answered that particular item. When *d* equals –1.00, then *none* of the students in the upper group but *all* of the students in the lower group correctly answered the particular question on the test. Both of these situations are very unlikely to occur. Most experts agree that an index of 0.30 or higher is generally acceptable.

There are various ways to group students' scores. The percentage of students' scores that you include in your analyses groupings depends on the number of scores that are in the total data set. The most common two ways include grouping scores either by the top and bottom 27% of scorers, which leaves 46% in the middle group. The middle group is not included in the analyses. The other way to group scores is to split all scores into only two groups of 50% in each. The number of test takers impacts how you select the method of grouping. Generally, large data sets use 27% of the top scores and 27% of the bottom scores. In a normal distribution of scores, the optimal point at which the groupings allow for discriminating among scores and maintaining high reliability or stability of results is at the 27% grouping (Kelly, 1939). However, if there is a small data set (low number of test takers), then it may be better to use 50%/50% groupings because 27% of the total may not be a large enough sample size to discriminate among scores.

To determine how well a particular test item discriminates between high and low scorers on the test, follow the following list of steps:

- **Step 1:** Order the tests from lowest to highest or highest to lowest.
- **Step 2:** Divide the tests into groups: upper group (27% of students who make the highest scores); lower group (27% of students who make the lowest scores);

middle group (remaining 46%). Use 50%/50% groupings if you have a low number of test takers (see example below). *Keep in mind, these are general percentages and some experts recommend other grouping strategies. If there is one remaining, add it to the upper group.*

- **Step 3:** Subtract the number of test-takers that correctly answered the item in the lower group from the number of test-takers that correctly answered the item in the upper group.
- **Step 4:** Divide the difference (number of lower group from the upper group) by the total number of test-takers in the upper group.

Let's walk through one example using two groups (each comprised of 50% of the total number of test takers).

Step 1: Order the tests from highest to lowest (or lowest to highest)

N = 21

Test μ (number correct) = 4.62

N upper group = 11

N lower group = 10

STUDENT ID	ITEM 1	ITEM 2	ITEM 3	ITEM 4	ITEM 5	ITEM 6	ITEM 7	TOTAL NUMBER CORRECT (BY STUDENT)
B	1	1	1	1	1	1	1	7
D	1	1	1	1	1	1	1	7
H	1	1	1	1	1	1	1	7
N	1	1	1	1	1	1	1	7
P	1	1	1	1	1	1	0	6
I	1	1	1	1	1	1	0	6
S	1	1	1	1	1	0	1	6
L	1	1	1	0	1	1	1	6
A	1	1	1	1	1	1	0	6
O	1	0	1	1	1	1	0	5
U	1	1	1	0	1	1	0	5
R	0	1	1	0	1	0	1	4
G	0	1	1	1	0	1	0	4

(Continued)

STUDENT ID	ITEM 1	ITEM 2	ITEM 3	ITEM 4	ITEM 5	ITEM 6	ITEM 7	TOTAL NUMBER CORRECT (BY STUDENT)
F	0	1	1	1	0	1	0	4
T	1	0	1	1	0	1	0	4
Q	1	1	1	1	0	0	0	4
M	0	1	1	0	1	0	0	3
C	0	1	1	0	1	0	0	3
K	1	1	1	0	0	0	0	3
J	0	0	1	0	0	0	0	1
E	0	0	0	0	0	0	0	0
Number correct	14	17	20	12	14	13	7	97
Total number that answered question (*n*)	21	21	21	21	21	21	21	147
Difficulty index (*p*)	0.667	0.810	0.952	0.571	0.667	0.619	0.333	0.660
Number in upper group that answered correctly	11	10	11	9	11	10	6	
Total number in upper group	11	11	11	11	11	11	11	
Number in lower group that answered correctly	3	7	9	4	3	3	1	
Total number in lower group	10	10	10	10	10	10	10	

Step 2: Divide total test-takers into groups (two groups given the small number of test-takers)—upper 50% and lower 50% signified by the boldface separation line in the table.

Step 3: Subtract the number of test-takers that correctly answered the item in the lower group from the number of test-takers that correctly answered the item in the upper group (see table below).

Step 4: Divide the difference (number of lower group from the upper group) by the total number of test-takers in the upper group (see table below).

The following table is a summarized format of the information contained in the example above. The table also includes each of the steps described above.

ITEM NUMBER	NUMBER CORRECT (UPPER GROUP)	NUMBER CORRECT (LOWER GROUP)	STEP 3 (DIFFERENCE IN UPPER AND LOWER NUMBER CORRECT)	DIFFICULTY INDEX	STEP 4 (DIFFERENCE IN NUMBER CORRECT DIVIDED BY NUMBER IN UPPER GROUP)	DISCRIMINATION INDEX (D)
1	11	3	8	0.667	8 ÷ 11	0.727
2	10	7	3	0.810	3 ÷ 11	0.273
3	11	9	2	0.952	2 ÷ 11	0.182
4	9	4	5	0.571	5 ÷ 11	0.455
5	11	3	8	0.667	8 ÷ 11	0.727
6	10	3	7	0.619	7 ÷ 11	0.636
7	6	1	5	0.333	5 ÷ 11	0.455

General Guidelines for Acceptable Discrimination Indices

The following table illustrates the generally agreed-upon guidelines for analyzing item discrimination.

$d \geq 0.40$	Very good item discrimination
$d = 0.30–0.39$	Good item discrimination
$d = 0.20–0.29$	Marginal item discrimination
$d \leq 0.19$	Poor item discrimination

Cautions Associated With Item Analyses. Item discrimination is directly related to item difficulty. Items with low- or high-difficulty indices will have low ability to discriminate between low and high scorers. If the test measures different constructs on a single instrument, item discrimination may be low because of the nature or style of the items themselves. Conducting a review of the item verbiage and response choices is recommended. A single analysis of test items is not recommended. These data are influenced by many factors including the number of test takers, student characteristics, instructional strategies used, and chance of error in the scores. Item analysis is not congruent with item or test validity. Instead, the analysis is a better indicator of internal consistency of the items.

WHAT DOES THE LITERATURE SAY ABOUT ITEM ANALYSIS?

Reading 7.1

ANALYZING TESTS

Malcolm L. Van Blerkom

INTRODUCTION

After you have given a test to your students and graded it, you are not necessarily finished with the test. You may choose to use the test or items from the test again sometime in the future. Therefore, it is useful to know how well the individual test items measure the students' knowledge of the material. Did the items work as you expected them to do? By analyzing the test, you can gain information on how well you taught the concepts to your students. Test items can tell you many things about student performance. Have you ever asked yourself any of these questions? Are there issues, ideas, and concepts that the students did not master? Are there issues about which the students appear to be confused? Is there material that will require further work? These questions can often be answered by performing a test analysis.

In recent years there has been more emphasis on using teaching techniques that are supported by research, that are data driven. When you complete a test analysis and use that information to alter your teaching and testing, you are using data-driven techniques.

TEST ANALYSIS

You can analyze your classroom tests at several levels. The first level of analysis involves examining the test as a whole. In an ideal world, when you give your students a summative assessment (a chapter or unit examination), you could find that every student obtains a perfect score on the exam. Assuming that your examination is highly valid and appropriately difficult, such a result would tell you that you have been very successful in teaching the material—that every student has successfully mastered the material. In reality, however, you rarely experience such an idealistic situation.

What does it actually mean when your students score very well on an examination? Let's say that the mean on the test was 95% correct with a very small standard deviation. Essentially, all students scored from about 90% to 100% correct. First, you should assume that you have a class that is, at least, somewhat heterogeneous—the students differ from one another in ability. So, how should you interpret such a performance?

There are, at least, two possible interpretations. The first interpretation is that the students may have all done a superb job in mastering the material. You could conclude that they all really knew what they were doing, and were able to demonstrate their skills on the test. However, there remains another possibility. That second possibility is that the test was much easier than you had anticipated. Are the scores much higher than the scores you typically obtain when testing similar students on the same material? Even students who had not fully mastered the material were able to perform well on the test simply because the items were very easy. In many instances, you should be able to rule out this second possibility by examining the individual test items to see how well they match your objectives. However, if you want further confirmation that the students truly mastered the material, you might wait a couple of weeks and then give the students another opportunity to demonstrate that they can actually use their learning in another type of assessment. If they again perform well, you can be more confident about your original interpretation of the data.[1]

Now, let's take a look at the other end of the continuum. What does it mean when students perform poorly on an assessment? One possibility is that you simply did not do a very good job of teaching the material, or that some of the material was more confusing to the students than you had anticipated. If you have not done enough formative assessment (discussed in Chapter 1), you may get a surprise when you give a summative assessment to find that the students perform poorly on some topics. There are some skills that the students simply have not mastered. Although you may be tempted to move on to the next chapter or unit (in order to stay on schedule), in most instances the responsible thing to do is to revisit the topics that confused the students. This is especially important if those topics with which the students are struggling include skills that they will need to use in the future.

When students perform poorly on a summative assessment, you also need to consider the possibility that the test was simply more difficult than you had anticipated. Again, you need to check to see that the test items match the objectives from that chapter or unit. Sometimes teachers give tests that are too difficult, or do not match what they taught. For example, some teachers provide instruction primarily at the knowledge and comprehension levels, but test primarily at the application and analysis levels. In this case, there is simply not a good match between the teaching and the testing. Their tests lack evidence of content-related validity because of this poor match between what was taught and what was tested.

ITEM ANALYSIS

The second level for evaluating a test is to perform an item analysis. Although you can perform an item analysis on any type of item, it is easiest to demonstrate with multiple-choice items. With any type of item you can examine item difficulty and item discrimination. However, with multiple-choice and alternative-choice items, you can look at three characteristics of each item. You can examine item difficulty, item discrimination, and you can perform a distractor analysis. Let us look at each of these separately.

Item Difficulty

Item difficulty is a rather straightforward concept—it is simply reported as the proportion of students who had the item correct. For example, if 73% of the students answered the item correctly, then the item difficulty is listed as .73. The only potentially confusing aspect of item "difficulty" is that its name is, in fact, the opposite—the higher the item difficulty, the easier the item. So, an item with a difficulty level of .88 is an easier item than one with a difficulty level of .55. More students could correctly answer the first item than the second.

Definition

Item difficulty is the proportion of students who answered the item correctly.

Why is item difficulty important? There are a variety of reasons why item difficulty could be important, but we are most frequently concerned with how item difficulty is related to item discrimination: A test item should differentiate between students who are well prepared for the exam and those students who are less prepared. The better-prepared students should be able to answer the item correctly more often than the less-prepared students. It turns out that, frequently, items that are moderately difficult better differentiate between well-prepared and less-prepared students than items that are either very difficult or items that are very easy. When items are very easy, most students are able to answer them correctly, including both the well-prepared and the less-prepared students. The same is true with very difficult items, except that most students get the items wrong; both the well-prepared and the less-prepared students are likely to answer them incorrectly. We often (but not always) find the greatest differences in performance between the better-prepared and less-prepared students when the items are moderately difficult. Therefore, you should strive to develop test items that are moderately difficult.

You might ask, what is a moderately difficult item? As I just mentioned, we frequently obtain the best item discrimination and the highest reliability on a test when the items are moderately difficult. But how difficult is moderately difficult? Frederic Lord (1952) demonstrated that, for completion items, we can obtain the best reliability and the greatest ability to differentiate when items have a difficulty level of about .50. However, for true–false and multiple-choice items, guessing can play a role and the items need to be somewhat easier in order to obtain the highest possible reliability. For example, with true–false items, since guessing can play a significant role, we tend to get the best discrimination and reliability when item difficulty is about .85. However, with five-option multiple-choice items, where the effect of guessing is less severe, we tend to get the best discrimination and the highest reliability when item difficulty is about .69.

Item Discrimination

As I just mentioned, a good item tends to discriminate (differentiate) between well-prepared and less-prepared students. You want more of the better-prepared students to get the item correct than the less-prepared students.

Figure 7.1 is an example of an item analysis for Item 17 from a 40-item four-option multiple-choice test that I gave to students in one of my psychology classes. For this item, option B was the correct choice (indicated by the asterisk). The first row of the item analysis shows the proportion of students from the upper quartile (the 25% of the students who had the highest total scores on the test) who selected each of the four options. None of the best-prepared students chose either options A or D, whereas, 80% of those students chose option B (the correct choice) and 20% chose option C. The next row of the item analysis shows the proportion of students from the lowest quartile (the 25% of the students who had the lowest total scores on the test) who selected each of the four options. None of the least-prepared students chose option D. Of this group, 50% chose option A, 20% chose option B (the correct choice), and 30% chose option C. The last row of the item analysis shows the proportion of all students from the class who chose each of the four options.

		ITEM 17		
STUDENTS	**A**	**B***	**C**	**D**
Upper 25%	0.00	0.80	0.20	0.00
Lower 25%	0.50	0.20	0.30	0.00
All students	0.36	0.38	0.26	0.00

FIGURE 7.1 **Item analysis for Item 17.**

How do we determine the item difficulty level? You will remember that the item difficulty is the proportion of students who had the item correct. The last row of Figure 7.1 shows that, among all of the students who took the test, 38% chose option B, the correct answer. Therefore, the item difficulty is listed as .38.

How do we determine item discrimination? The *item discrimination index* is the difference in the proportion of the better-prepared students who had the item correct as compared to the proportion of less-prepared students who had the item correct. In this case 80% (or .80) of the best-prepared students had the item correct, whereas only 20% (or .20) of the least-prepared students had the item correct. The item discrimination index is computed by subtracting the second proportion from the first (Equation 7.1).

$$\text{Item Discrimination Index} = .80 - .20 = .60 \tag{7.1}$$

Essentially this means that 60% more of the better-prepared students got this item correct than the less-prepared students.

Definition

The **item discrimination index** is the proportion of the better-prepared students who had the item correct minus the proportion of less-prepared students who had the item correct.

You should strive to build tests with high reliability. When selecting test items, you should attempt to use items with higher item discrimination indices because item discrimination contributes to a test's reliability. A test that has many items with a high item discrimination index will have a higher reliability than a test which has many items with a lower item discrimination index.

How high should an item discrimination index be? It could conceivably range from −1.00 to 1.00. An item discrimination index of .40 or higher indicates that it is an exceptionally good item in terms of its ability to discriminate between well-prepared and less-prepared students. Item discrimination indices in the range of .20 to .39 are very desirable and help the reliability of the test. When item discrimination indices are near zero (.00), the item is neither helping nor harming the reliability of the test. However, negative item discrimination indices (yes, they do happen) actually lower the reliability of the test.

Perhaps the most common reason to perform an item analysis is to examine a test that you might want to use again in the future. The item analysis allows you to determine which items are working well and should be retained, which should perhaps be discarded, and which could benefit from some revision. An ideal item is one that is moderately difficult and has a high discrimination index. The emphasis should be on the discrimination index. If this is low or even negative, the item needs to be revised or even discarded. You need to determine, or at least make a good guess at, why the item is performing poorly. Is there an error in the answer key? Does the item include material that was not covered because of time constraints or other circumstances? Is the item poorly worded and potentially confusing? When you use multiple-choice items you can take one additional step in the item analysis: You can also perform a distractor analysis.

Distractor Analysis

With multiple-choice items, the well-prepared students are expected to be able to determine that any distractor (a wrong alternative) is clearly the wrong answer. However, students who are less prepared should view each distractor as potentially correct. So, how should a distractor behave? Essentially, you would like to see a higher proportion of the less-prepared students choose a distractor than the well-prepared students.

In truth, we do not typically complete a distractor analysis unless an item is performing poorly. Let's once again look at Figure 7.1. Option A appears to be working very well as a distractor since none of the best-prepared students chose it and 50% of the less-prepared students chose it. Option C is also working, although not quite as well since 20% of the best-prepared students chose it and 30% of the less-prepared students did. Option D is not working at all since none of the students chose it. Since option D is not working, you might be tempted to change it. However, since this item has an item discrimination index of .60 and is working extremely well, it would be best to simply not touch Option D. Sometimes you simply need to leave well-enough alone!

Before you receive additional practice with item analysis, let's discuss the typical steps you should go through to perform an analysis. Remember, you are primarily trying to find items which discriminate well. Step 1: Does this item display an acceptable item discrimination index? If the answer is "yes," you could stop with the item analysis right there. If the answer is "no," then you should move to Step 2 and look at the item difficulty level. Is the item discrimination index too low simply because the item is either too easy or too difficult? If so, then you might want to revise the item simply to make it easier or more difficult. However, if the item discriminates poorly, yet is moderately difficult, then you should move on to Step 3 and perform a distractor analysis. Are the distractors working as they should?

How can distractors work poorly? One way in which distractors fail to perform is that few, if any, students choose them. These distractors are often implausible. Even the least-prepared students are able to recognize them as clearly wrong.

Another potential problem with distractors is where they are partially correct. These can be confusing because even the well-prepared students can sometimes be attracted to them. If any of these problems occur, rewriting the distractor can sometimes improve the question.

Some other distractors, such as "all of the above" or "none of the above," are potentially troublesome. For example, some students who have chosen the correct alternative but are not 100% certain that they are correct may be inclined to go with "none of the above" simply because of their uncertainty. If the "none of the above" alternative were not present, the students would have had the answer correct as they should have. Also, if students identify one alternative as clearly correct and one or more other alternatives as partially correct, they may be inclined to go for "all of the above" as the correct answer. In many situations, this type of alternative simply makes items more confusing than they should be and tends to lower the overall reliability of a test.

Item Analysis Practice

Now that we have gone through the process of performing an item analysis, let us look at a few item analyses to get some practice.

Figure 7.2 represents the item analyses of five additional items from the psychology test mentioned earlier. Let's start with **Item 19**. Start by looking at the item's discrimination index. The correct alternative is A. Of the best-prepared students, .70 chose that option, whereas only .30 of the least-prepared students chose A. The difference between .70 and .30 is .40, which is the item discrimination index. Since this is an excellent item discrimination level, you could stop the item analysis right here. However, since you are practicing, we will continue with the other steps.

The second step is to check the item difficulty. Since 54% of all of the students chose option A, the correct answer, the item difficulty is listed as .54. The high item discrimination index is likely to be attributable, in part, to the fact that this is a moderately difficult item.

The third step is the distractor analysis. Option B is not working well as a distractor since only 10% of all the students chose it and none of either the better-prepared or the less-prepared students chose it. Option C is working well since more of the less-prepared students chose it than the better-prepared students. However, 30% of the best-prepared students chose it, which is a cause for concern. Perhaps it appeared to be more correct than it should have done. Option D is not working well either since only one student out of the class chose it. For this item, even though none of the distractors are working as well as they could, the item, as a whole, is still working well. (Sometimes you just

ITEM 19				
STUDENTS	A*	B	C	D
Upper 25%	0.70	0.00	0.30	0.00
Lower 25%	0.30	0.10	0.50	0.10
All students	0.54	0.10	0.33	0.03

ITEM 26				
STUDENTS	A	B	C*	D
Upper 25%	0.20	0.20	0.60	0.00
Lower 25%	0.40	0.30	0.30	0.00
All students	0.23	0.18	0.56	0.03

ITEM 32				
STUDENTS	A	B	C	D*
Upper 25%	0.00	0.00	0.00	1.00
Lower 25%	0.10	0.00	0.20	0.70
All students	0.05	0.00	0.13	0.82

ITEM 35				
STUDENTS	A*	B	C	D
Upper 25%	0.80	0.10	0.00	0.10
Lower 25%	0.70	0.00	0.10	0.20
All students	0.77	0.03	0.13	0.08

ITEM 44				
STUDENTS	A	B*	C	D
Upper 25%	0.15	0.25	0.50	0.15
Lower 25%	0.10	0.40	0.20	0.30
All students	0.12	0.30	0.40	0.18

FIGURE 7.2 **Additional item analyses.**

get lucky!) In that case, I would probably leave the item as is, but check it again the next time I use it.

For **Item 26**, the correct option is C. Here, .60 of the best-prepared students chose option C, but only .30 of the least-prepared students chose that option, making the item discrimination index .60 minus .30, or .30. Therefore, this item has an acceptable item discrimination index. The difficulty index is also acceptable since 56% of all of the students in the class chose the correct answer.

The distractor analysis indicates that both options A and B are working well as distractors since more of the less-prepared students chose each. However, since none of either the better-prepared or the less-prepared students chose option D, that distractor is simply not doing well. Perhaps you could improve this item by making option D more plausible.

The next item analysis is for **Item 32**. All of the better-prepared students chose option D (the correct option) and 70% of the less-prepared students chose it. Therefore, the item discrimination index is 1.00 minus .70, or .30. This item also has an acceptable item discrimination index.

This item was relatively easy since, of all the students in the class, 82% chose the correct answer. Since this was a relatively easy item, you might be surprised that it still had an acceptable item discrimination index. You may recall that we tend to see better item discrimination with items that are moderately difficult. However, once again, as Frederic Lord (1952) pointed out, with multiple-choice items guessing can play a role. Therefore, he argued, with four-option multiple choice items, you typically obtain the best item discrimination and the best reliability for items with a difficulty index of about .74. Although Item 32 is not very difficult, its difficulty index does not fall far from the ideal difficulty level which allows good item discrimination.

Let's look at the distractor analysis. For this item, option A is somewhat helpful because more of the less-prepared students chose it, but it is not contributing significantly to the item's effectiveness. Option B is not working at all since no student chose it. Finally, option C is helping since none of the best-prepared students chose it and 20% of the less-prepared students did. If you wanted to improve this item, you should work on making options A and B more plausible.

Item 35 is an interesting item since, at first, it appears to have characteristics similar to the previous item. However, it has a very low item discrimination index because, even though 80% of the best-prepared students had the item correct, 70% of the least-prepared students also had it correct. The item discrimination index is, therefore, only .10. Why is it performing so poorly? This item, with an item difficulty of .77, is slightly more difficult than the last item we examined and, therefore, could still potentially work well. However, since it is not working well we need to look at the distractors.

Both options C and D are each helping the item somewhat since more of the less-prepared students chose them. However, option B is actually hurting the item because it was more attractive to the well-prepared students than it was to the less-prepared students.

To improve this item, you should probably start by trying to make it a little more difficult. Then, closely examine each of the distractors to attempt to make each of them more plausible.

Finally, let's look at **Item 44**. The correct option B was answered by 40% of the less-prepared students, but by only 25% of the best-prepared students. Therefore, this item actually has a negative item discrimination index of −.15. Clearly, there is something wrong with this item. The item was quite difficult since only 30% of all of the students chose the correct option. Why is this item so difficult?

The distractor analysis immediately points to the possible problem with this item. A total of 40% of the students chose option C and 50% of the better-prepared students chose this option. Why is this option so attractive, especially to the better-prepared students? When an item performs like this, you should first check to see if it was keyed incorrectly. Perhaps option C is actually the correct answer. If it turns out that option C was the correct answer, then the mystery is solved.

If the item was keyed correctly and option B was the correct answer, then you must look at the actual item. What was it about option C that made it appear to be a better choice even to the better-prepared students? You may be able to see the problem and change that option. However, at other times you may have to resort to asking the class why they thought that option C was correct. Frequently, they are able to explain how they interpreted the question which made option C appear to be correct. Generally, it has to do with the confusing way the question was phrased. In those instances, rewriting the question stem to make it crystal clear can help. However, if you cannot determine why the question was performing so poorly, it is often better simply to throw out the question and replace it with another question the next time you use that exam.

The Stability of Item Analyses

At this point I should add a word of caution about test analyses and item analyses. Both types of analysis will depend on the students who took the test. For example, I frequently teach two or more sections of a particular course each term. If I give the same test to each section and perform separate analyses on each test, I can get different results. I may find that Item 26 worked well in one section of the class, but rather poorly in the other section. Differences from section to section are more likely if the class sizes are small. If I had 18 students in one section and 22 students in the other section, then I should expect differences in the analyses. However, if the class sizes were much larger (for example, 75 students in each section), then I would expect more similarities between the separate analyses. Therefore, I recommend that you exercise caution in interpreting test and item analyses with very small classes.

SUMMARY

After students complete a test you may choose to perform a test analysis. This allows you to examine how the test performed as an assessment device and sometimes gives you information about how well you taught the unit. Since your goal is to use tests and other assessment devices that are reliable, the students who are the most knowledgeable and the best prepared should earn high scores, whereas those students who are less prepared should earn lower scores.

If you would like to use the test again in the future, you might want to perform an item analysis to determine which items can be used again, which items require revision, and which should be replaced. For almost any type of test item, you can examine the item difficulty and the item discrimination index. In addition, for multiple-choice items you can also perform a distractor analysis.

EXERCISES

The following charts show the results from a 6th-grade world geography exam. (Note: The correct answer for each item is noted with an asterisk.)

Analyze each item using the following steps:

a. Find the item discrimination index.

b. Find the item difficulty.

c. Perform a distractor analysis, when appropriate to do so.

d. Explain what you would change about the item, if anything, and why.

ITEM 1				
STUDENTS	**A**	**B**	**C***	**D**
Upper 25%	0.25	0.35	0.40	0.00
Lower 25%	0.40	0.10	0.50	0.00
All	0.30	0.20	0.45	0.05

ITEM 2				
STUDENTS	A*	B	C	D
Upper 25%	0.75	0.05	0.15	0.05
Lower 25%	0.73	0.00	0.20	0.07
All	0.74	0.02	0.18	0.06

ITEM 3				
STUDENTS	A*	B	C	D
Upper 25%	0.52	0.00	0.00	0.48
Lower 25%	0.16	0.00	0.20	0.64
All	0.45	0.03	0.05	0.47

ITEM 4				
STUDENTS	A	B	C	D*
Upper 25%	0.02	0.00	0.05	0.93
Lower 25%	0.10	0.29	0.45	0.16
All	0.05	0.15	0.30	0.50

SPOTLIGHT ON THE CLASSROOM

Mr. Eswein teaches 6th-grade social studies at Fairmont Middle School near Birmingham, Alabama. He has taught the same class for over 10 years, and has altered his tests, as well as his teaching methods, many times. In the past he relied on items from the teacher's edition of the textbook to build his exams. However, this year he has decided that he could build better exam items on his own. His tests typically consist of multiple-choice, true-false, and short essay questions.

This year, his students appear to be having more trouble with his tests and are performing poorly, even though they appear to be equally as talented as his students have been for the past several years. They eagerly participate in class discussions and give every appearance of understanding the material. When he uses various informal assessments in the classroom, his students appear able to understand the material, but still perform poorly on his chapter and unit examinations.

He is now beginning to wonder if his exam items are the source of the problem.

Can you suggest some steps that he could take to evaluate his tests and his exam items?

NOTE

1 When you give a test immediately after instruction, students will tend to perform well simply because the material is fresh in their memory. However, if your assessment is delayed by several weeks, you can expect a decline in the students' performance as a result of forgetting the material. Many educators argue that a delayed test is actually a better indicator of real learning. After all, most of the time we need to use skills learned some time ago.

REFERENCE

Lord, F. M. (1952). The relation of the reliability of multiple-choice tests to the distribution of item difficulty. *Psychometrika, 17* (2), 181–194.

HOW DOES ITEM ANALYSIS APPLY TO MY PROFESSION?

The key to successful item analysis is to look at all data for each item. When making decisions about which items to retain and which to omit, you should make your decisions after the completion of a comprehensive analysis. There are some general guidelines to follow for difficulty and discrimination indices. But, ultimately, the decision is yours. The stakes associated with the test should also be considered when making decisions about which items to retain and which to omit.

As an educator, item analysis is one method of determining the quality of test items included on tests. The procedures for calculating indices of difficulty and discrimination are fairly straightforward and can be quickly accomplished in a classroom setting. Although the results of item analyses will not provide you with a complete picture of quality, they do provide you with some information related to the quality of the test and its items. Thus, analyzing items is something that I encourage you to do every time you administer a test. You will quickly identify which items are good measures of students' knowledge and/or skills and which are not. If the results are lower than recommended guidelines, do not use those items to determine consequences for students (grades, retention/promotion, ability groupings, etc.). Remember, it is our responsibility to do no harm to our students. Making decisions and using results elicited from faulty items can do just that. Therefore, at a minimum, review the items and reword them prior to using them again to test students' knowledge and/or skills.

CHAPTER 8

Performance Assessment in Traditionally Hard-to-Measure Disciplines

What and how do I assess?

AS YOU READ

As you read the chapter, think about the following:

- How do **you** define or explain performance assessment?
- What is **your** experience with performance assessment?

WHAT ARE THE GENERAL PRINCIPLES OF PERFORMANCE ASSESSMENT IN TRADITIONALLY HARD-TO-MEASURE DISCIPLINES?

In an era of accountability that incorporates growth modeling and value added, what should be included in disciplines such as these that are difficult to measure? Performance assessment is often used in these disciplines. **Performance assessment** refers to measuring the demonstration of a skill. But, in some disciplines, the skill is only one aspect of what should be measured. Teachers and leaders should also consider knowledge and affect in these disciplines as well.

There are some academic disciplines that are much more difficult to determine what we should assess in our students. For example, if you think about assessing students in art, music, and/or physical education—how do you determine students' grades? Do you base the grade on their skill level? What about their knowledge of the content? What about their attitude toward that content? Well, there is no one correct answer. However, I have done quite a bit of research in this area and I strongly believe that the best approach is to use a model that incorporates all three: knowledge, skill, and affect. As is the case in other disciplines, teachers' time is valuable. Using a three-foci assessment approach is often more time consuming than simply giving students an exam to assess their knowledge. There is also controversy around using students' level of skill to assess them in these disciplines. To what extent do students' skills be taken into consideration? These are all things that educators must decide. Sometimes, these are individual teacher's decisions. Others, these types of decisions are schoolwide. In other cases, these decisions are made by the district or state level to try to have consistency across schools. The bottom line is there will be varying levels of skill among students. There will be varying levels of attitudes among students. And, there will be varying levels of knowledge among students. To assess each of those constructs, there are multiple approaches.

Assessing Knowledge in Hard-to-Measure Disciplines

This book has already discussed the challenges to assessing students' knowledge in any discipline. For example, educators need to consider the items themselves, the type of assessment tool to use, varying assessments to meet the needs of all students, not over-assessing the students, ensuring the questions align to the content, and so on. That is no different in these traditionally hard-to-measure disciplines. Educators have the same considerations when assessing physical education, art, or music, just as they do when assessing science knowledge. Although assessing knowledge is often thought of as "easier" than assessing other factors, it still isn't easy as you learned in previous chapters.

Assessing Skills in Hard-to-Measure Disciplines

There is some controversy about how to best assess students' skills in art, music, and physical education. There is disagreement on what skills should be assessed in what grade levels. There are multiple curriculum standards that guide these types of decisions; however, there is little guidance to what level the student should demonstrate his/her skills. For example, do you focus on the mechanics of the skill themselves? Do you focus on the outcome/result? Let's look at an example of throwing a ball. Here is a hypothetical 4th grade physical education standard that states, "Throw objects of various sizes and shapes to a stationary individual from multiple distances using a correct overhand motion." As the teacher, you conduct multiple lessons that instruct students on the proper way to throw balls of various sizes and shapes. You teach them the proper form, strength necessary for specified distances, and so forth. The students have time to practice their throwing skills and now it comes time to assess their skills. So, you decide to partner the students and have them throw balls of various sizes and shapes to a stationary individual at multiple distances. Student 1 is right on target with her throws. She throws each of three balls to her partner and five different distances. She uses proper form in all instances. Student 2 throws the same balls to her partner but can't quite throw it far enough to reach her partner at the longest distance. She demonstrates proper form but lacks the strength. Student 3 throws the same balls to his partner and overthrows the partner by 10 feet. Proper form is used; however, too much strength is used that results in overthrowing his partner. Student 4 can't throw the heaviest ball even halfway to his partner. She does not use the proper form and instead uses underhand throwing. In this scenario, all four students demonstrated differing levels of skill for this particular standard. How do you determine which students meet the standard and which do not? At what level of skill must they demonstrate their ability? Some students simply have more ability than others in the discipline. For these reasons, some disciplines are more difficult to assess than others.

Assessing Affect in Hard-to-Measure Disciplines

Student affect is often the forgotten attribute in testing and measurement. Affect can be an important aspect of the "whole child." Specifically, dedication toward a particular concept or discipline may be an important factor for assessment purposes. Think about teaching health and physical education. You teach about various concepts and skills but the really important part of that discipline is healthy living. You can't really assess healthy living in a classroom setting. So, perhaps you incorporate journals and logs and attitudes about healthy living into your comprehensive assessment plan. By doing so, however, you have tremendous difficulty accurately and fairly measuring student's documents. So, perhaps you use an attitude inventory to evaluate one's attitude toward living.

WHAT DOES THE LITERATURE SAY ABOUT PERFORMANCE ASSESSMENT IN TRADITIONALLY HARD-TO-MEASURE DISCIPLINES?

Reading 8.1

PERFORMANCE ASSESSMENT IN CTE

Bart Washer and Lori Cochran

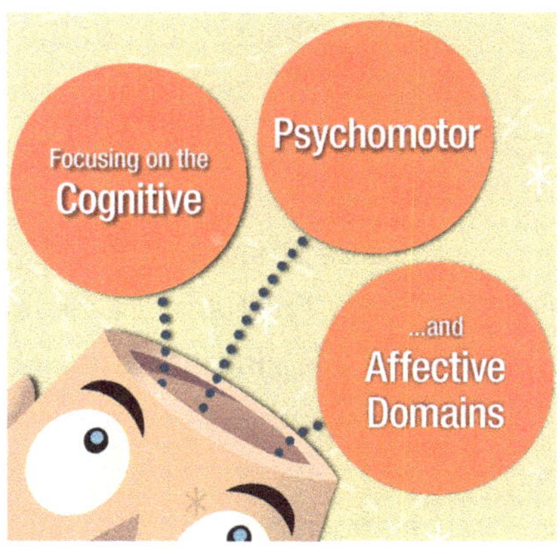

If we told you Eddie Haskell from the TV show "Leave it to Beaver," high school driver's ed courses, and educational theory from 1956 would drive this article, would you stay with us? After all, many of us may have never heard of Eddie Haskell, taken driver's ed, nor were around in 1956. But if you'll give us 20 minutes (plus reflection time), we believe you'll gain a greater appreciation of assessing and documenting CTE students' performances in the cognitive, psychomotor and affective domains.

BACKGROUND TERMINOLOGY AND CONTEXT

Throughout this article we will be referring to curriculum alignment, direct/indirect/process/product assessments (Erickson & Wentling, 1988), authentic assessments (Custer, Schell, McAlister, Scott, & Hoepfl, 2000), Assessment FOR Learning (Stiggins, 2004), and student performance in the three learning domains as identified by Bloom, Engelhart, Furst, Hill, and Krathwohl, (1956). Rather than formally defining these terms, we'll present examples in a brief vignette from a course many of us took in high school: driver's education.

DRIVER'S EDUCATION WITH MR. BRADY

Let's call our driver's ed (DE) instructor Mr. Brady. One of the activities in DE was to change an automobile tire according to the manufacturer's recommendation (learner objective). Mr. Brady communicated the objective to the class, used the appropriate instructional resources (the DE car, tools, parking lot), taught us to change a tire (instructional strategy), and then *planned* on assessing us using the manufacturer's instructions for changing the tire. Mr. Brady planned to assess and document our *processes* (procedural steps) along the way rather than only assessing the final work *product*. Up to this point, this activity had curriculum alignment. The objectives, resources, instructional strategies and planned assessments were connected.

But then Mr. Brady realized the dynamic of our class during this activity and how much instructional time we could waste if we really applied ourselves. Short on time, he was forced to modify his assessment. Rather than each of us actually changing a tire (psychomotor assessment) for our summative exam (which would have been a direct assessment of the objective), Mr. Brady revised the assessment where we explained how to change a tire on a written test (cognitive assessment of the psychomotor objective). So while Mr. Brady was not *directly assessing* the psychomotor objective, he was *indirectly assessing* our knowledge of the psychomotor *process*.

While *indirect* assessment may jeopardize curriculum alignment, Mr. Brady realized that *indirectly* assessing the psychomotor objective trumped no assessment. Furthermore, he required us to perform on the cognitive assessment by synthesizing our knowledge and practicing correct grammar (using short answer and essay questions) rather than using lower-level matching, true/false (T/F), and multiple choice (M/C) items. Finally, by observing us throughout the instructional process using *Assessment FOR Learning*-type practices (Stiggins,

2004), he was providing formative feedback throughout instruction while documenting our performance. Even if we didn't have our psychomotor test on actually changing the tire, he had documentation (formative assessment data) on what every student knew, could do and valued.

INVESTIGATING THE THREE DOMAINS OF LEARNING

Using the above tire-changing vignette, it is proposed that we (Mr. Brady's students) were *performing* in all three learning domains: cognitive (knowledge), psychomotor (fine/gross motor skills), and affective (appreciations, values) (Bloom et al, 1956; Krathwohl, Bloom & Masia, 1964). We also propose performance is not tied only to the psychomotor domain. Upon reflection, don't our students perform in all domains almost daily? Perhaps they do not perform in the psychomotor domain every day, but we believe students perform cognitive and affective tasks daily. We must seize the opportunity in documenting student performance in all three learning domains as evidence of their learning.

For example in the cognitive domain, students perform daily by using correct English and grammar when they create memos, resumes, technical documents, charts and solutions to problems, etc. In the psychomotor domain, they perform fine and gross motor skills by completing physical tasks like welding, constructing a structure, and manipulating a mouse/pointing device. In the affective domain, they perform employability skills daily as evidenced by being on time, getting along with others, and spending time on task. But back to Mr. Brady's tire-changing activity: While we performed in all three domains, were Mr. Brady's assessments *authentic* (Custer et al., 2000)? Did his assessment mirror the tasks we would be performing in the real world when we would graduate?

AUTHENTIC STUDENT PERFORMANCE IN ALL THREE LEARNING DOMAINS

When determining the authenticity of our assessments, Custer, et al (2000) recommend this litmus test: Will our students be evaluated this way in industry? We further suggest investigating Custer's recommendation at the domain level. Cognitively, will students take recognition-response-based T/F, matching and M/C tests? Perhaps M/C items are relevant as many certification exams use scenario-based M/C items. But we believe that if a student can *construct* an answer (constructed-response items), they can recognize an answer (recognition-response items). Psychomotorly, will students be expected to perform the tasks of their craft with fine motor skills or gross motor skills? From an affective/employability perspective, are we preparing model employees? We may answer "yes" to these questions, but what do authentic assessments and the resulting documentation actually look like?

Regarding documentation, when authentically assessing student performance, we recommend two main types: student-completed and rater-completed (where the instructor, peer or guest rater observes and documents student performance). While not exclusive, the cognitive domain may generally include student-completed items while the psychomotor/affective domains generally include rater-completed items.

The Cognitive Domain

Many inservice and preservice teachers indicate using recognition-response items that are easy to grade (T/F, M/C, matching). While we must value teacher time (haven't we all created assessments with ease-of-grading in mind?), we also must ensure students are authentically performing in the cognitive domain that mirrors how they will perform in industry. When they gain employment, will they be describing and explaining procedures to a patient's family members? Will they be charting or journaling occurrences of a daily shift or customer repair order? Or will they be taking T/F, M/C, and matching tests? While recognition-response items may have their place for informal daily checks of understanding and reading quizzes, we propose student cognitive performance in higher-level assessments (chapter exams and above) should mirror how students will perform in their careers, using as many writing and public speaking activities as possible.

The Psychomotor Domain

Documenting student performance in the psychomotor domain (direct assessment) may shift from student-completed to rater-completed items. Whether the assessment is an actual project being constructed, a fictional job sheet, or a collaborative capstone project, Erickson and Wentling (1988) recommend using Likert-type rating scales (1–5); graphic rating scales (pictures or text that describe each Likert rating digit); a binary checklist (either yes or no); or organized, well-written notes. Rubrics (grading grids) with multiple criteria may also be used. Regardless of which tools we use, all items' criteria should be objective for rater consistency.

The Affective Domain

We encourage using daily checklists to document student performance and behaviors in the affective domain, which also include employability skills. Using checklists, students earn points (we don't encourage giving or taking away student points) in areas such as arriving ready to learn, getting along well with others, following all safety procedures, staying on task, and not leaving early. In an oversimplified example, if students earn one point daily for each employability criteria listed above (five points total per day), over a typical 175-day school year, students can earn up to 875 points. If a student is late every day (unexcused tardies) but earns all other daily points, they would earn 700 daily points (80% average). This documentation of employability skills could justify a student's grade during parent-teacher conferences, but it must be communicated in the course syllabus. Yes, this requires dedicated documentation, but do we and our advisory committees believe employability skills are important? Absolutely. Moreover, as Custer and Claiborne (1992, 1995) discovered, employers value students' employability skills even more than we as educators think they do. So isn't it equally important for us to assess and document these required affective and employability skills?

However, the affective domain can be difficult to measure because we cannot see inside our students. Sometimes we only see what the students want us to see, which brings us to Eddie Haskell and The Eddie Haskell Syndrome. Eddie was the tormentor of Theodore "Beaver" Cleaver and a friend of Wally (Beaver's older brother) in the television show "Leave It to Beaver." Eddie always seemed to harass Beaver and was considered by many to be the "dark angel" of Wally's conscience. However, when Ward or June Cleaver (Beaver's and Wally's parents) were around, Eddie was a model citizen. Comments like, "My, what a lovely dress, Mrs. Cleaver!" would transition to "Hey Wally, where'd your old lady find that ugly dress?" when Mrs. Cleaver would leave the room. So cognitively, Eddie would tell adults what he thought they wanted to hear. But affectively, his true beliefs and values came out when he thought the adults weren't present; hence the term The Eddie Haskell Syndrome (EHS).

While it's difficult to accurately cite the origin of EHS, we've used it with our Missouri preservice and inservice teachers since 1995. Students like Eddie may cognitively perform by writing essays on topics such as "Why safety and personal protective equipment (PPE) are important." Using only the cognitive domain, we may measure and evaluate Eddie's knowledge and determine he values safety because he wrote it correctly in an essay. However, if we could observe unobtrusively, we may see Eddie chiding other students for wearing their PPE while not wearing his own. So what would that tell us about his true appreciations for safety and PPE? Sure, Eddie may have scored well on the cognitive assessment (he knows the importance), but does he truly appreciate the importance, based on his actions when he thinks we're not observing him?

Our challenge is this: If we ask our students affective questions on a cognitive (written) assessment (thus indirectly assessing the affective domain), they may tell us what they think we want to hear or properly recite our expectations. But how do they act when they think we're not observing them?

FINAL THOUGHTS

Student performance assessment can occur within the entire instructional process and should not be reserved for only the cognitive domain or summative assessments, nor occur only on test days. So to conclude, here are questions and challenges that hopefully summarize our main points.

1. How can we create (or revise as needed) our learner objectives to include all three learning domains? When hired, our students may be evaluated daily in all three domains, so shouldn't we provide the most thorough preparation possible?
2. How do we ensure curriculum alignment between our objectives, instructional strategies, resources and assessments, as all elements equally contribute to student learning?
3. How can we adapt (or increase) Assessment for Learning strategies that document student performance in all three domains, not waiting until test day or only using assessment of student learning?
4. How can we ensure our students will authentically perform in the cognitive domain as they will in their careers? When hired, will they be recognizing answers from a list of potential answers, or will they be constructing responses via written and spoken words?
5. In the affective and psychomotor domains, how can we improve or develop new tools (binary checklists, rating scales, graphic rating scales, rubrics) that document student performance throughout the instructional day? Furthermore, how can we incorporate process assessments where formative feedback is needed, and product assessment where there could be multiple safe and equally-efficient correct solutions?

Finally, we hope you were asking yourself while you were reading this if all three domains actually work together when assessing CTE students. We believe they do. For example, when a student is performing in the psychomotor domain, we believe the student is also performing in the cognitive domain (sequencing steps, evaluating the situation) and in the affective domain (appreciating a job well done, quality control, safety).

As Dabney Doty, former instructor at the University of Central Missouri accurately stated, "There is more headwork than handwork in shopwork." So we believe the key for CTE instructors is authentically assessing and documenting students' performances in all three learning domains and providing appropriate feedback to increase their learning.

Bart Washer, Ph.D., is an associate professor and CTE graduate program coordinator in the Dept. of Career and Technology Education at the University of Central Missouri. He may be reached at bwasher@ucmo.edu.

Lori Cochran is an assistant professor and coordinator of the Bachelor of Science degree in occupational education in the Dept. of Career and Technology Education at the University of Central Missouri. She may be reached at cochran@ucmo.edu.

REFERENCES

Bloom, B. S., Engelhart, M. D., Furst, E. J., Hill, W. H. & Krathwohl, D. R. (1956). *Taxonomy of Educational Objectives; The Classification of Educational Goals. Handbook I: Cognitive Domain.* New York, NY: David McKay Company, Inc.

Custer, R.L (Ed), Schell, J.W., McAlister, B., Scott, J. & Hoepfl, M. (2000). "Using Authentic Assessment in Vocational Education." Information Series No. 381. ERIC Clearing House on Adult, Career, and Vocational Education. Columbus, OH: The Ohio State University. Obtained from www.eric.ed.gov

Custer, R.L. & Claiborne, D.M. (1992). Critical skills cluster for vocational education. *Journal of Vocational Education Research*, 17(4), 15–40.

Custer, R.L., & Claiborne, D.M. (1995). "Critical Skills Cluster for Vocational Education: The Employers' Perspective: A Replication Study. *Journal of Vocational Education Research*, 20(1), 7–33.

Erickson, R.C., and Wentling, T.L. (1988). *Measuring Sudent Growth: Techniques and Procedures for Occupational Education* (Revised edition). Urbana, IL: Griffon Press.

Krathwohl, D.R., Bloom, B.S., & Masia, B.B. (1964). *Taxonomy of Educational Objectives; The Classification of Educational Goals. Handbook II: The Affective Domain.* New York, NY: David McKay Company, Inc.

Stiggins, R. (2004). "New Assessment Beliefs for a New School Mission." *Phi Delta Kappan*, 86(1), 22–27.

HOW DOES PERFORMANCE ASSESSMENT IN TRADITIONALLY HARD-TO-MEASURE DISCIPLINES APPLY TO MY PROFESSION?

You may be thinking that you don't teach a discipline that is labeled as hard-to-measure. However, as described throughout this chapter, you can see that performance assessment can be applied in multiple disciplines. As a current or future leader, assessing student and teacher affect may be an important component to better understanding the classroom and school culture. Furthermore, there are multiple educational policies that impact teachers' performance and potentially their pay. As a current or future educational leader, it is imperative that you develop systems to determine a fair and consistent manner in which you evaluate teacher performance. If student results on standardized tests are part of that evaluation process, what will you do to include or not include those that teach content areas for which there are no standardized tests? It is important that we think of ways to effectively and appropriately assess students regardless of the academic discipline.

CHAPTER 9

Assigning Grades: Homework, Classwork, and Exams

How much weight should these have?

AS YOU READ

As you read the chapter, think about the following:

- How do **you** define or explain grading?
- What is **your** experience with grading?

WHAT ARE THE GENERAL PRINCIPLES OF ASSIGNING GRADES?

There are multiple considerations when assigning grades to students. On what do you base those decisions? Do you consider students' effort? Do you use scores from only summative assessments to determine grades? Do you include homework in the final grades? What about ability, attitude, participation, and behavior? Should those variables impact student grades? Do student grades represent student achievement? According to Linn and Miller (2005), "In the final analysis, letter grades should reflect the extent to which students have achieved the learning outcomes specified in the course objectives, and these should be weighted according to their relative importance" (p. 377). They suggest that grades should be based on student achievement alone and that other variables should not be considered. In fact, multiple experts in the field of testing and measurement agree that final grades should represent student achievement or how well they performed on assessments that measured their knowledge of the curriculum standards. There is the belief that if you consider other factors, grades are then not consistent across students and they are not fair to all. Also, most parents believe that grades represent student achievement. They equate high grades with high achievement. When teachers consider other variables in the final grades, there is no way to know how much of the grade actually represents achievement.

There are counterarguments to the claims made in the paragraph above. Although the majority of measurement experts recommend considering only student achievement in final grades, the majority of practicing teachers use additional factors when assigning grades. Studies have confirmed that although student achievement is the most important factor in determining grades, other factors such as effort (measured by participation and homework completion) are also considered (Stiggins, Frisbie, & Griswold, 1989; Cross & Frary, 1999). Randall and Engelhard (2010) found that low-achieving and low-ability students that behave in class and put forth a tremendous amount of effort are rewarded with their final grade.

So, what should you do? The majority of measurement experts recommend that you use only student achievement to represent final grades. Yet, most classroom teachers, regardless of that recommendation, use other factors when determining final grades of students. Perhaps there are fair and consistent ways to incorporate achievement while taking some other factors into consideration. The table below lists some things to consider when assigning final grades.

CONSIDERATIONS	PROS	CONS
Comparisons with other students	Outstanding academic performance (comparatively speaking) is rewarded. This allows even the low-achievers to feel success.	Some students' grades may be lowered and others may be increased based on the overall performance of the group.
Comparisons to standards	The focus is on achieving course goals/objectives and not on trying to "get an A."	Educators have to clearly define course goals/objectives and the level of expectations for all students needs to be reasonable and consistent. The same students are typically rewarded/punished based on their academic achievements. Low achievers may experience learned-helplessness.
Comparisons to improvement	Educational expectations can be personalized and based on prior knowledge of students.	Growth over a short time period may be hard to measure. Students with more prior knowledge will show less growth than those with little prior knowledge.

Source: Adapted from https://citl.illinois.edu/citl-101/measurement-evaluation/exam-scoring/assigning-course-grades

WHAT DOES THE LITERATURE SAY ABOUT ASSIGNING GRADES?

Reading 9.1

VOTING FOR SUBJECTIVITY: ADDING SOME GRAY AREAS TO BLACK-AND-WHITE, OBJECTIVE GRADING PRACTICES

Jennifer L. Stitt and Judith J. Pula

The authors argue that teachers should include subjectivity as well as objectivity in their grading processes. Rewarding such intangibles as effort, attitude, participation, and improvement can help counterbalance grades lowered by test anxiety, for instance, and show that teachers care—and, in so doing, feed students' drive to succeed. The authors offer steps for adding or reinstating subjectivity into classroom grading practices.

In a society where education focuses on objective learning standards and assessments, teachers are often baffled about whether or not they should include lesson plans, assessment tools, or grading criteria that are not stringently linked to objectivity. Educators should not forget, however, that the gray areas of subjectivity—effort, attitude, participation, improvement, and so on—are linked to the reactions, personalities, and attitudes that reflect the core of human nature: empathy. Although society demands students to become adults who contribute via skills measured through objective standards, relying solely on objectivity seems to forsake a part of humans' drive to succeed: the empathetic side driven by emotion. Ultimately, teachers are left to ask themselves whether to be only subjective, only objective, or both. Given the purpose of education in relation to the betterment of society and a student's eventual participation in that society, teachers should impart knowledge, use grade assessment tools, and communicate student performance both objectively and subjectively.

THE GRADING DIVIDE

Arguably, the root of what may be called a *grading divide* does not lie with today's teachers. Because some educators may lack the professional training and development appropriate for implementing their school's grading systems, "most [teachers] simply replicate what they experienced as students" (Guskey, Swan, & Jung, 2011, p. 53) when they were in grade school. This replication becomes an immediate problem because teachers employed together in the same school likely have varied educational experiences. These experiences impact their perspectives about what should and should not be considered for grades and rarely "reflect those recommended by researchers and align ... with a standards-based approach" (Guskey, Swan, & Jung, p. 53). Furthermore, students are not the only ones negatively affected by such widely varied beliefs about and approaches to grading. Inconsistency traps teachers as well, especially when the time comes to defend why a certain student receives a particular grade. In essence, teachers' perceptions are sometimes the only support offered for grades that are supposed to communicate reflective observation of student progress.

To complicate the grading divide further, educational reforms of the last decade make it difficult for teachers to justify subjective grading. Demands for standardized learning control lesson plans and classroom activities, which leads to an equal demand for measurable methods of imparting knowledge and improving student performance. Logically, such performance can be measured most easily through use of assessment tools that are equally balanced and aligned with standards. This standards-driven system intends to set students up for success rather than failure by removing "the possibility for subjectivity at every link in the educational chain" (Narter, 2005, p. 65). In reality, however, complete removal of subjectivity would reduce humans to teaching and learning automatons, eliminating room for the kind of empathetic communication and development that motivate learners. In a similar way, society may believe that being "objective would prevent biases" (Malouff, 2008, p. 191), but working to prevent biases does not always completely eliminate them. If the betterment of students so they can successfully contribute to society as adults is a key goal of education, then teachers and students should cultivate a classroom that includes both subjective and objective assessment—a classroom where biases can be minimized because both types of assessment can be used as checks and balances for one another.

IMPLEMENTING SUBJECTIVITY APPROPRIATELY

Consequently, if educators agree that communication of human understanding is important to the process of education, then subjectivity must rightfully play a role alongside objectivity in grading students' performances. Teachers can mold students to be active, rule-abiding, expectation-meeting adults in society by requiring them to be accountable for meeting certain requirements even under what some would consider the more subjective categories of grading ("Should student attitudes," 2002). Some examples of these categories might include

- a *participation grade* for contributing to discussions, measured through the number of times the student participates within a given time frame;

- a *completion grade* for responsibility in completion and retention of long-term assignments, which pairs with an objective grade based on a rubric;
- a *preparation grade* for coming prepared for performance with all materials necessary, as well as an objective grade for any assignments that were due.

Jennifer L. Stitt is a recent graduate of Frostburg State University's Master of Arts in Teaching program. She earned a Bachelor of Science in English (literature) with a minor in theater at the same institution. She is an alumnae member and former president of the campus's chapter of Sigma Tau Delta, the international English honors society. Stitt also worked as a tutor at the campus writing center, where she gained Level II CRLA certification. She currently teaches senior English and creative writing at Hampshire High School in Romney, West Virginia. JLBryStitt@gmail.com

Judith J. Pula, PhD, a member and past president of the Alpha Alpha Chapter (MD), specializes in teaching composition and reading, particularly to underprepared students, at Frostburg State University in Western Maryland. She believes the freshman-year experience lays a foundation crucial to success in college and beyond, so she strives to bring first-year students to a realization of themselves as self-empowered, independent writers and members of the academic community. jpula@frostburg.edu

In contrast, teachers should not determine grades based on areas that are not supported easily or that are not relevant to learning, such as a student's being visually attractive, performing at a certain level in previous classes under their instruction, being of a particular gender or ethnicity, seeming to be of a certain intellectual or socioeconomic level, or having some sort of connection to the teacher outside of the classroom setting (Malouff, 2008). Although grading based on any of the above criteria can occur either consciously or subconsciously inside grading parameters, teachers need to be wary to implement consistent definition and application of grading criteria in both objective and subjective realms in order to defend their grading decisions properly.

Steps. Teachers can begin balancing subjectivity and objectivity in their classroom in a few steps. The first step is precisely defining categories related to the existing structure of the classroom's learning and working environments.

Teachers are typically aware of the workload they expect students to complete and the conditions under which they wish students to meet learning goals, so the decision of what matters in the way of subjective-grading categories should make sense when paired with objective-grading categories. For example, if a teacher prefers lecturing methods, then students should not be expected to meet the requirements of a participation grade—any more than a classroom fueled by cooperative learning should be without a participation grade.

The second step requires teachers to track students' progress on subjective measures over longer periods, just as objective measures require day-to-day, assignment-to-assignment consideration. In working with subjective categories, teachers must additionally be aware that some students have issues related to elements outside of the classroom affecting their performances, such as emotional distress from home, in different ways over different periods of time. Accordingly, teachers must be patient in developing an understanding of the duration of varied issues and how to deal with them in the classroom ("Should student attitudes," 2002). Teachers who take the time to understand and know what their students feel and think will quickly be able to distinguish among these issues and the length of their effects.

The third step to proper implementation of subjective grading is determining the percentage that subjectivity counts toward the total grade at the end of a grading period. Subjectively-graded categories should not be major, deciding factors in the grading process; rather, they should serve as support for the grades students have earned through objective measures. Assessment in subjective categories can allow students a chance at bettering grades when their application of knowledge fails due to faults that are not totally under their control. For instance, an effort category worth 10% of a total grade can save a student who has testing anxiety from failing the whole marking period. Of course, this example only works if a teacher tends to favor tests over other assessments such as papers, lab reports, or projects. Hence, the first step of grading subjectively comes into play again: subjective categories should be in alignment with the objective tools being used. With this cycle of steps, teachers are more likely to find success in creating a properly-supported grading system that uses objectivity and subjectivity to balance and check each other.

By using these steps to incorporate subjectivity alongside objective-grading methods already being used, teachers become more than simply vehicles to guide learning; teachers retain their humanity and indicate that they care. A classroom without such humanity is not a classroom.

Some people may argue that subjectively grading students in addition to grading them objectively will do nothing to improve their performance. On the contrary, we concur with the claim that "students can tell if [a teacher] really care[s] ... and will work harder on their behavior and studies for a teacher who does" ("Should student attitudes," 2002, p. 16). Additionally, a grade based solely on performance under objective standards does little to inform parents about their child's behavior as a student. The aforementioned student who has test anxieties might be mistaken as lazy at home for achieving poor grades. In the classroom, this child may be a teacher's most diligent pupil, wanting more knowledge and more chances to improve. Therefore, grades in subjective areas provide a bonus: communication between parents and teachers opens further to provide an accurate picture of how the student performs and to stimulate progressive conversation about the student as a whole person. The importance of communication on this topic cannot be forgotten, given that education is not the responsibility of teachers alone. As Narter (2005) insightfully observed, "Education is a dynamic undertaking between human beings within a common culture in which teachers impart knowledge by means of careful selection and concern for" (p. 65) students and parents alike. Meaningful communication around both objective and subjective criteria more fully informs parents in this dynamic undertaking.

CONCLUSION

Grades based purely on objective criteria leave little room for communication about subjective goals in the classroom that rely on the empathetic characteristics of human nature, and few teachers can prove full objectivity in conscious and subconscious decisions 100% of the time. Accordingly, subjective aspects of grading, teaching, and learning should be addressed instead of simply cast aside. Overlooking these aspects forgets the empathetic core that drives humans to be successful for their own sakes and for the sake of society. If teachers are purely objective in the way they pass on knowledge and assess students, then they should not be surprised if their students are not empathetic to others. Teachers are supposed to be a reflection of the society that the students will join as adults, and blending objectivity and subjectivity will help communicate what well-rounded students should be once they become contributing adults in society.

REFERENCES

Guskey, T. R., Swan, G. M., & Jung, L. A. (2011). Grades that mean something: Kentucky develops standards-based report cards. *Phi Delta Kappan*, *93*(2), 52–57. Retrieved from http://search.proquest.com.proxy-fs.researchport.umd.edu/docview/896432056?accountid=27669

Malouff, J. (2008). Bias in grading. *College Teaching*, 56(3), 191–192. Retrieved from http://search.proquest.com/docview/274714538?accountid=27669

Narter, D. (2005). Teacher as machine: The cost of objectivity. *English Journal, 84*(4), 65–69. Retrieved from http://search.proquest.com.proxy-fs.researchport.umd.edu/docview/237295889?accountid=27669

Should student attitudes and behaviors be a part of the grading process in K–12 classes? (2013). Issues. *JOPERD—The Journal of Physical Education, Recreation & Dance*, 73(9), 12–17. doi:10.1080/07303084.2002.10608339

HOW DOES ASSIGNING GRADES APPLY TO MY PROFESSION?

Assigning grades has significant implications for students, teachers, and administrators. There are no straightforward and/or easy answers to the best approach to assigning grades. This chapter provided some considerations for teachers. But also, it is highly recommended that educational leaders think about consistency across courses and/or disciplines. However, it is imperative that groups of individuals have a voice in making such decisions. It is not only necessary to receive feedback from multiple stakeholders, but also, these policies are reviewed on an ongoing basis. Every class of students is different. What works with one class may not work with another class. Teachers know their students better than anyone else in the educational system, so it is important that the system is fair for all students.

CHAPTER 10

Test Scores and Data Interpretation

Data and statistics: What do they mean?

AS YOU READ

As you read the chapter think about the following:

- How do **you** define or explain test scores and interpreting data?
- What is **your** experience with test scores and interpreting data?

WHAT ARE THE GENERAL PRINCIPLES OF INTERPRETING DATA?

Scales of Measurement

Scales of measurement are ways in which we identify/define/characterize variables and/or numbers. There are multiple scales of measurement. In this chapter, we will focus on the four most common: nominal, ordinal, interval, and ratio. These are the names that we give to different types of data so that we can make decisions based on those data.

Nominal. This is the name given to categorical data used simply as an identifier. For example, the number on the back of your child's baseball jersey. Is there any numeric meaning for that particular number? I'm not asking if there is sentimental meaning—I'm asking if there is numeric meaning. Absolutely not, these numbers simply represent or identify a particular child. What if you wanted to determine if boys score differently than girls on a particular assessment? Well, to make this determination, you have to identify "boys" and "girls." Typically, we use the numbers 1 and 2 or 0 and 1 to "code" the children into a particular category. Boys are in category "1" and girls are in category "2." The numbers are insignificant—they are simply identifiers so we can analyze data and make inferences from the data about the students.

Ordinal. This is the name given to data that represent a series of relationships or a rank. For example, if you have a spelling bee contest, the child that correctly spells the most words comes in first place. The next comes in second place, the next in third, etc. Does this mean that the first place ranked student correctly spelled exactly one word more than the one in second, and two more than the one in third, and three more than the one in fourth, etc.? Not necessarily. A key component to identifying data as ordinal is that the distance between the ranks is not the same. A good example is when you think of running a timed race. The first-place finisher might be 10 minutes ahead of the second-place finisher who might be 30 seconds ahead of the third-place finisher. The scales themselves do not represent a measurable quantity—we simply "rank" order the items/people/contestants, etc. Likert scales are actually ordinal scales; however, we typically use them as interval (I'll explain below).

Interval. This is the name given to data that are continuous and that have equal distances between points (unlike ordinal). A common example is temperature. Think about measuring the temperature of water on the stove as it eventually comes to a boil. You insert the thermometer and watch the temperature increase as heat is applied. After watching the temperature rise from 80 degrees Fahrenheit, then 81, then 82, 83, etc. all the way until it reaches a boil at 212 degrees Fahrenheit. Think about this: Is 212 degrees Fahrenheit twice as hot as 106 degrees Fahrenheit? You might think it is; but, "hotness" is not actually measurable—remember above when we discussed assigning properties to a latent variable in order to actually measure it? Well, degrees Fahrenheit, although commonly accepted in the United States is the way in which we measure temperature, it's actually just an assigned descriptor; therefore, 212 is *not* necessarily twice as hot as 106. Interval data does not necessarily have an absolute zero. If it did, it would be called something more specific (hint: it is next). It is also for this reason that Likert scales are *not* actually interval; but rather, they are ordinal. We often use Likert scales as a measurement of attitudes, beliefs, feelings, etc. These are all latent constructs (variables). When we ask someone to provide a level of agreement versus disagreement, or satisfaction versus dissatisfaction about a given topic/concept; we are asking for their opinion. Opinions don't have the exact same distance between the points. Meaning, when we are disgusted by something, we may feel very strongly about that disgusting thing. When we are ecstatic about something, we may feel very strongly about that thing. We may actually feel stronger about one thing than another or another or another. Each item on the questionnaire may actually elicit various emotions for which we rank our opinions (or level of emotion, satisfaction, excitement, etc.). As such, the distances between the numeric values placed on the scales are almost completely meaningless (because they're not the same in a given setting, across groups, etc.). Because the distances between points are not the same, Likert scales are ordinal. We often treat them as interval because we need at least interval data to make inferences from samples about populations.

Ratio. This is the name given to data that has an absolute zero. That is, if one quantity is twice that of another, then the number does represent double. Example: if you are 6 feet tall and your friend is 3 feet tall—you are twice as tall as your friend. Height and weight are by far the most common illustrations of this type of data. Other examples include centimeters, meters, inches, etc. There are no such things as negative lengths.

Measures of Central Tendency

Central tendency. The central tendency is a generic term describing the center of a frequency distribution of observed scores as measured by the mean, median, and mode.

Mean. The mean is often thought of as the average of the scores (although technically the mode, median, and mean are all averages). Simply add up each score and divide by the total number of scores. The mean is sensitive to every single value within the data set. Thus, if two students scored a 2 out of 100 possible points, these two scores can pull down the overall mean—which will skew the distribution of scores. If there are several low scores, the tail of the distribution will point to the right (positively skewed distribution). If there are several high scores, the tail of the distribution will point to the left (negatively skewed distribution). See figure 10.1.

Median. The 50th percentile is the point that divides the distribution of scores into two equal halves. The median is often the best for descriptive purposes. Think about how median income is often discussed instead of the mean income of Americans. To calculate the median, you align all of the scores in order (highest to lowest or lowest to highest) then find the exact middle. This middle might be between two numbers in the data set or it might be one of the data points. Look at the following examples and then determine the median.

- 1, 2, 3, 4, 5, 6: The median is 3.5 (it's the center point). If there is an even amount of numbers in the data set, then the median will fall between the two center numbers.
- 1, 2, 3, 4, 5, 6, 7: The median is 4 (it's the center point). If there is an odd amount of numbers in the data set, then the middle number will be the median.

Mode. The mode is the score that is the most frequent in the data set. The mode is not very stable (it is not a good way to report the results of an assessment). For assessment purposes, it does not provide much information. In a data set with the following numbers: 1, 2, 2, 3, 3, 4, 4, 5, 5, 6, 6, 6, 6—the mode is 6 because it is the number that appears most frequently. What about this example? 1, 2, 2, 3, 3, 4, 5, 6—the mode is 2 and 3 (this distribution of scores is bimodal—it has two bumps on a graph instead of one).

Normal distribution. The normal distribution is a distribution of scores that is perfectly symmetrical. It has a skewness and kurtosis of zero. This is sometimes referred to as the bell curve.

Skewed distribution. A skewed distribution is when the tail of the distribution points to the left or the right.

Kurtosis. Kurtosis refers to the height of the peak of the data.

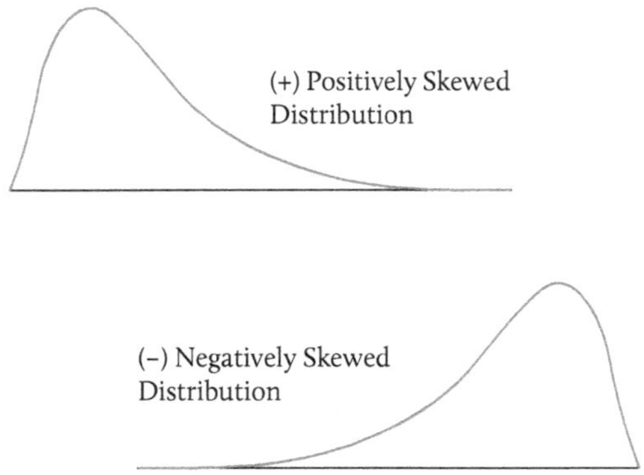

FIGURE 10.1

Measures of Dispersion and Relative Position

Measures of dispersion. This refers to the measure of the extent to which scores in a data set are spread out or differ from each other. The most common measures of dispersion are variance, range, and standard deviation.

Range. This is simply the term that is used to explain the difference between the highest and the lowest scores in the distribution. Simply stated, the range of scores (provided below) is 67 to 98 *or* 31.

Variance. This is the term used to describe the average squared difference of scores from the mean score of a distribution. Let's walk through an example using test scores. Below is the data set for a small group of students:

Student scores: 67, 86, 89, 90, 98

To calculate the variance, we first need to determine the mean of the scores for this particular sample of students. Calculate the mean (add up the numbers and divide by the total number of data points). The mean is approximately 86. Now we have to subtract each number from the mean. This is called deviation from the mean. (See below.)

$$86 - 67 = 19$$
$$86 - 86 = 0$$
$$86 - 89 = -3$$
$$86 - 90 = -4$$
$$86 - 98 = -12$$

To verify the accuracy of your numbers, at this point in time, you could simply add up the numbers. If the sum of all of the deviations from the mean add up to 0; then you're correct so far!

Now, to obtain the variance, you need to square each of your numbers (otherwise you'd come up with a variance of 0 which isn't what we are looking for). So, square each number (the differences you obtained above) (19, 0, –3, –4, –12) then add up the total.

$$361 + 9 + 16 + 144 = 530$$

The final step is to divide the total by the number of scores (if you are calculating the population variance) or one less than the number of scores/the number minus one (if you are calculating the sample variance). Let's assume that in this case, the five scores we have represent a sample of students so you should calculate the sample variance by dividing by 4, which is (n – 1). Where n = the total number of scores/participants/items/etc. So, in this case, there are a total number of five scores. Subtract 1, which means that you'll divide by 4 (instead of 5). If you had scores for the entire population, then you'd divide by N (number in the population) instead of n – 1 (sample).

530 divided by 4 scores = 132.5
Variance = 132.5

Standard deviation. This is a very, very important concept. In general, the standard deviation (SD) describes the spread of scores around the mean. To calculate the standard deviation, you simply take the square root of the variance. In this case, the standard deviation is 11.5. A SD of 11.5 seems fairly large for a group of five scores. This one piece of information informs you that there is a large spread of scores (they are widely dispersed).

Measures of relative position. Unlike measures of dispersion, some scores measure a specific location in the data set. Standard scores (which are also called z-scores), quartile scores, and percentile scores are common measures of relative position.

Percentiles. This is the term used to describe the relative position of a score or where in the distribution a score falls. It's all relative. A score falls into a certain percentile that is equal to or above that percent of scores. If your child scored in the 98th percentile, then he scored higher than or equal to 98% of the other students or he scored lower than 2% of the other students. Remember, the median, the middle score, is the 50th percentile.

Quartile. This is another measure of relative position. There are four quartiles in the entire set of data points. Simply take the scores and split them into four groups of 25% of the scores (or fourths). Scores that fall into the first quartile are within the lowest 25% of scores. Scores that fall into the fourth quartile are higher than at least 75% of the scores.

z-scores. This is another useful measure of relative position and is often called a standard score. These are standardized scores that are calculated using the mean (a measure of central tendency) and the standard deviation (a measure of dispersion). They are used to determine if relative differences between data points are large or small, above or below their respective means. When assessment results are compared, the teacher often uses raw scores to make these comparisons. Well, sometimes it's difficult to make such comparisons (especially if the raw scores have little or no meaning). Take for example a test that has scores that range from 100 to 1,000. The scores themselves have little meaning, so we standardize them so we can make statements that make sense about the raw scores. An even better example is if you want to compare scores from two different tests (different raw scores). You can convert raw scores into z-scores for comparison purposes. Essentially, the *z*-score is a measure of standard deviation units that the raw score is from the mean so it is relative. Because it is relative, we can make sense of the relationships between scores. Let's begin with an example by using the sample of scores from above (where the mean is 86 and SD is 11.5). Determine the *z*-score for a raw score of 95. To do this, you simply plug in the numbers into the formula below.

$$z = \frac{(X - \overline{x})}{s}$$

$$z = \frac{(95 - 86)}{11.5} = .78$$

z = z-score
X = observed score
$\overline{x}$ = sample mean
s = sample standard deviation

As you may have noticed, the example used sample statistics because the scores were compiled from only a sample of test scores. For classroom teachers, this is the most common scenario; however, if you know the scores of an entire population of students that completed a test, you would use population statistics instead of sample statistics. Instead of using the sample mean ($\overline{x}$), use the population mean (μ), and use the population standard deviation (σ) instead of the sample standard deviation (s). See the formula below:

$$z = \frac{X - \mu}{\sigma}$$

z = z-score
X = observed score

μ = population mean
σ = population standard deviation

If the z-score is positive, then the raw score is above the mean. If it is negative, then the raw score is below the mean. You may be wondering why we need to know about z-scores. Well, we do because it's imperative that you know how to interpret percentiles and convert raw scores into percentiles. To make these conversions, you need to understand z-scores.

Other Types of Scores

Raw scores and observed scores. Typically, the raw score and the observed score are the same. A raw score is the number correct that the student receives. Observed scores are the number or percentage correct that the student receives. These scores are sometimes more meaningful than other times. For example, raw and/or observed scores are what teachers record in their grade books when calculating grades based on teacher-created assessments for students in their classes. Raw/observed scores are typically manipulated in some manner when scores are reported for performance on standardized tests. These raw/observed scores are typically entered into a sophisticated statistical formula that converts the raw/observed scores into some established scale score. These scale scores are then either reported back to the teachers and parents *or* the scale scores are then again converted to a corresponding level or rating scale.

Scaled scores. Scaled scores are often used in reporting standardized test results. They are the result obtained when some statistical transformation is applied to the raw scores to make scores across examinees consistent over time. These transformed scores are used to report results using a consistent scale for multiple test takers and multiple years.

True Scores, Error, Standard Error of Measurement, and Confidence Intervals

Recall from Chapter 3, a true score is the unaffected score (constant throughout the series of measurements) that is not possible to obtain in an educational setting. If someone took a test an infinite number of times, then eventually a true score could be targeted. Think of it in terms of an average score of a test that a student took an infinite number of times.

In the realm of testing, there is always some error. Error can positively or negatively affect the score on a test. What if the same test was given multiple times? It is possible that the student/person changed from one testing to the next. For example, she could have received special training/coaching from a tutor. This would likely change her knowledge base. Perhaps she was extremely motivated one day. This level of motivation could change her result on the test. Another form of error is that the test could be slightly different between measurements. If this is the case, it is called a parallel form of the test. This is when the teacher thinks she is measuring the same content as the first test; however, the questions might be worded slightly different than the first administration. For some students, they may do better on the second set of questions. For others, they may perform worse. Finally, the limited sample of tests (number of repetitions) may add to the error of a test. For example, it is likely that error will be higher if the test is only taken one or two times. There are multiple reasons for this: perhaps Suzie fell asleep for a few minutes during the test, or perhaps Johnny wouldn't stop tapping his foot and it bothered another student. If students were given multiple times to take a test (which has other issues, such as the "carry-over effects") then the likelihood of error is decreased, and you can obtain a more accurate measure of a student's true score.

Thus, true score is the observed score (the score the student receives on the test) plus the error (regardless of the origin of the error). Although we say *plus* error, you use the error (plus or minus) to create a range of scores to determine the likelihood that the student's **true score** is within a particular range. This range is titled a **confidence interval** because we are not calculating actual error; but, in reality we calculate the standard error of measurement (SEM). We simply use this SEM to determine the likelihood that a child's true score will fall within a particular range. Again, this is called a confidence interval.

To determine a confidence interval, you must calculate the SEM. Think of SEM as being similar to standard deviations. SEMs are essentially the standard deviation of the variations in the test scores. This means that if you took 200 readings of your weight on a bathroom scale, you could produce a frequency distribution of your weights. A frequency distribution includes the number of times each weight is derived. For example, the following table shows a possible frequency distribution for someone's weight if one took 200 measurements of his weight over a period of time.

WEIGHT IN LBS.	FREQUENCY (N = 200)
125	25
126	45
127	40
128	40
129	30
130	20

To determine the SEM, first determine the average value of the frequency distribution (127.325 lbs.). You can think of this average value as the "true" weight. This distribution also has a standard deviation that describes the spread of the measurements around

administrators use data such as the mean of scores to compare the performance of a particular test-taker with that of the other examinees. However, the mean does not always provide us with a thorough summary of the results. As you may have noticed, in the above example there is only one low score of 27, which is known as the **outlier**, but it considerably lowers the mean and gives the wrong impression of the overall performance of the group. To achieve a more accurate outline of the scores we also need to know the dispersion or the spread of the scores. The most commonly used measures of dispersion are the **range** and the **standard deviation** (SD).

To indicate the range we simply give the highest and the lowest scores. In other words, the range is the distance between the lowest and the highest score. If we use the above example, the range of scores would be 27–85. Once again, we find this data somewhat misleading since there is essentially only one score below 30 in these series.

The SD is the most dependable index of dispersion. It represents the degree of the deviation of the scores from the mean and gives us a complete picture of how widely the scores spread around the mean. A small standard deviation indicates that the scores are closely clustered together; a large standard deviation is a sign of a more or less even spread of scores over a broad range. Therefore, the larger the standard deviation, the more variable the set of scores.

Let us look at the scores of a small group of seven test-takers as an example.

The mean for this group is 49 and the SD is 2.1, which is indicative of an average range of scores. The following is the formula for calculating the SD.

$$SD = \sqrt{\frac{\sum(X - \bar{X})^2}{n - 1}}$$

Where *X* is the individual test-taker's score:

$\bar{X}$ = the mean

n = the number of test-takers.

$$SD = \sqrt{\frac{46}{7 - 1}} = \sqrt{4.667} = 2.16$$

The SD is rarely a round figure, but in order to simplify the discussion, we will round down the SD for this group to 2. Procedures for calculating the SD through SPSS will be described in the next section, along with other statistical computations by SPSS.

It should be pointed out however, that the SD gets its clear-cut meaning when the scores have a **normal** distribution. This can be defined as a set of scores that are spread evenly above and below the mean, similar to the set of scores given

TABLE 10.1 Example of scores for a small group of test-takers

STUDENT	SCORE
1	46
2	47
3	48
4	49
5	50
6	51
7	52

above. This form of distribution is common in educational test scores. The normal distribution is demonstrated by a symmetrical curve having a bell-like shape.

Let us assume that the scores in a given test are symmetrically distributed. As shown in Figure 10.1, there are usually three SD units below the mean and three SD units above the mean. The percentages of the scores that fall in each standard deviation interval are as follows:

- 68.2% of the scores will fall within one SD above and one SD below the mean.
- 95.4% of the scores will fall within two SDs above and two SDs below the mean.
- 99.8% will fall within three SDs above and three SDs below the mean.

How do these statistics help us in the overall assessments of students? To answer this question I will re-examine the example of the group of seven test-takers given above. As suggested before, the SD of 2 in this group of scores tells us that scores are distributed more or less evenly over a broad range; they are not bunched together around the mean. In general, tests that are designed to measure various levels of proficiency, such as English placement tests, should have a relatively high standard deviation since variability is essential in these forms of tests. It allows the teachers or administrators to group the students into a few levels of instruction based on their score. On the other hand, in the case of achievement tests, where the purpose of the test is to determine who will pass or fail in a particular course, a wide spread of scores is not essential.

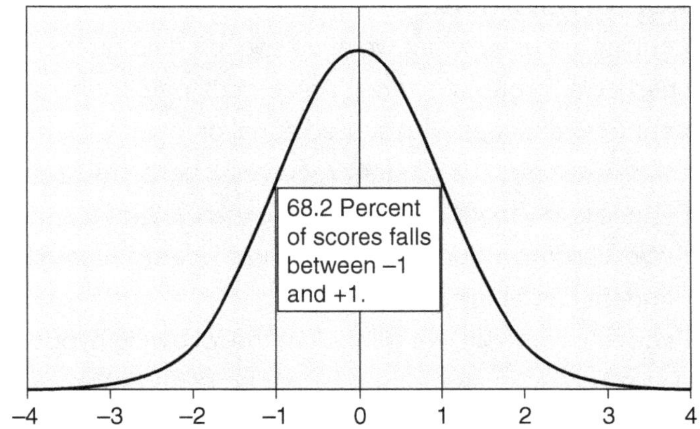

FIGURE 10.2 **Normal probability curve**

In addition, the standard deviation index in a normal distribution gives us a much better basis than raw scores to evaluate the status of an individual in a given group. Let us look at test-taker number 6 in Table 10.1. The score of 51 places him in one standard deviation above the mean. As demonstrated in Figure 10.2, one standard deviation above the mean is above 82% of the other scores. So we can conclude that the student number 6 has done quite well.

CALCULATIONS WITH SPSS

As pointed out earlier, a simple method of computing statistical concepts is through SPSS, which is a relatively new, straightforward program. Follow these steps to calculate descriptive statistics which include the mean and the SD. These instructions are designed for Windows 12.1. Other programs may not be identical but similar.

- Click on SPSS icon.
- Double-click the gray box in column 1 entitled "var0001."
- Give your variable a name.
- Enter the scores under variable 1.
- Go to the top of the page and select "Analyze."
- Go to the menu and click on "descriptive statistics."
- Select "descriptives."
- Move the variable you want to analyze to the box by the arrow.
- Click on "OK."

The mean and the standard deviation will appear on the next page.

PERCENTILE RANKS

Percentiles are used when we need specific data on the ranking of a test-taker in a particular group of scores. A subject's percentile rank is indicative of the percent of examinees who scored equal or lower than he or she did in a given test. For example, if a student receives a percentile rank of 92 in the test of TOEFL, we know that he scored higher than 92% of the students who took that test. We also know that only 8% of the students scored higher than him, thus the student has a high standing.

To give an example, we will look at the following hypothetical band of scores for a small group of students in higher percentile ranks.

Table 10.2 is an example of a band of scores for students in higher percentile ranks.

TABLE 10.2 **Example of band of scores for students in high percentile rank**

SCORES	PERCENTILE RANKS
96–100	99
92–95	97
84–91	93
79–83	88

It should be pointed out, however, that percentiles are mainly associated with norm-referenced tests rather than criterion-referenced tests, as discussed in Chapter 1.

STANDARDIZED SCORES

Another method of representing a student's relative position in a group is to convert his or her raw score, which is the number of correct answers, into standardized score. As Brown (1996) argues, teachers of English are mainly familiar with raw scores and they find standardized scores somewhat confusing. For example, for a long time many ESL teachers and administrators had difficulty interpreting the standardized score reported by TOEFL. It was hard to figure out a standardized score such as 53 in reading comprehension when there were only 30 items on the test of reading. The following is a brief explanation on the procedures used by TOEFL and other standardized tests to convert raw scores into standard scores.

Standardized scores convey test performance in terms of standard deviation units from the mean. There are several types of standard scores, but they all use, more or less, the same approach. I will therefore, discuss the most common types: the z and the T scores. The following is the formula for calculating the z score.

$$Z - score = \frac{X - \overline{X}}{SD}$$

Where:

X = any raw score
$\overline{X}$ = mean of raw scores
SD = standard deviation of raw scores.

We can easily apply this formula to any raw score. For example, the z score for the raw score of 51 in Table 10.1 would be calculated as follows:

$$Z - score = \frac{51 - 49}{2} = +1.0$$

The z score of the student number 9 is therefore +1, which is one standard deviation above the mean.

The z score, though very well known, has some limitations. The main problem with z scores is that they can be plus or minus depending on the raw score. The *z*-score is always minus when the raw score is smaller than the mean. Not including the minus can lead to major confusions in interpreting the test results. The major advantages of norm-referenced tests is that they are usually well established, especially in respect to such matters such as item analysis and estimation of reliability (Hughes, 2003).

The z score is usually transformed into a T score, which is reported in positive numbers. To obtain T scores we arbitrarily multiply the *z*-score by 10 and add the results to 50. Therefore,

T score = 10*(z) + 50

For z of +1 the T score will be calculated as the following:

T = 10(1) + 50 = 60

Since standardized scores are the [principal] method of reporting scores in standardized tests, it is important for both ESL teachers and administrators to have some basic knowledge of the concept.

CORRELATION

One of most beneficial areas of statistical analysis in education is correlation.

The main purpose of correlation analysis in language assessment is to examine the degree of association between two sets of scores or variables. Determining the degree of correlation is central to the discussion of language testing since it will prepare the readers for a better understanding of reliability and validity [...]. For example, we might need to examine the relationship between two set of scores on two tests taken by the same group of students (reliability) or the relationship between the TOEFL score acquired by a group of students and their actual performance in writing (validity). There is certainly some degree of correlation between these variables; the question is how great. The degree to which two variables are associated is estimated by calculating the coefficient of correlation, symbolized by the letter *r*. It should be noted that if the correlation is positive, the highest value of *r* is 1. However, if the correlation is negative, the lowest value is –1. On the other hand, if there is no correlation, the value of r is 0. When there is a perfect positive relationship the value is 1. Let us consider the following example, where we look at the scores of a group of 10 students taking two similar tests. Let us hypothetically say that the first test is midterm and the second test is the final. We would like to know the correlation between the midterm and the final exam for this group. See Table 10.3.

TABLE 10.3 Example of students' scores for correlation

STUDENT	MID-TERM	FINAL
A	X	Y
B	24	26
C	28	30
D	30	32
E	34	38
F	40	42
G	50	52
H	52	53
I	60	62
J	62	64
K	64	66

A correlation coefficient of 0.99 in this example shows a perfect correlation between these two tests.

The following is the formula for computing the correlation coefficient:

$$r = \frac{SS_{xy}}{\sqrt{SS_x SS_y}}$$

where

$$SS_{xy} = \sum xy - \frac{(\sum x)(\sum y)}{n}$$

$$SS_x = \sum x^2 - \frac{(\sum x)^2}{n}, and$$

$$SS_y = \sum y^2 - \frac{(\sum y)^2}{n}$$

This formula appears complex, but it really involves simple arithmetic. While it is important to be able to understand and implement this formula in the computation of basic statistical concepts, I would like to encourage the learners to use the SPSS program, which has simplified these calculations.

The following are the steps to follow to calculate the Pearson method of correlation coefficient by SPSS:

- Click on the SPSS icon.
- Enter your scores under X and Y in the two columns of variable 1 and variable 2.
- Go to the statistics menu.
- Select "correlate" and "bivariate."
- Move the variables to the variable window.
- Make sure Pearson box is checked.
- Press "OK."

The correlation coefficient will appear on the next page.

The correlation coefficient can also be negative, as previously explained. Negative correlation is indicative of two variables having a negative correlation or being exactly the opposite in the highest value of *r* at –1.

As mentioned earlier, correlation is a critical aspect of statistical analysis. In addition to its widespread use in determining test reliability and validity, correlation analysis is used for research in education and language acquisition. For example, the relationship between reading and vocabulary knowledge or the correlation between performance skills such as speaking and writing are some of current topics of research.

CONCLUSION

In this chapter I have presented a brief explanation of a number of statistical concepts that are central to the discussion of language testing. There are some additional statistical procedures, such as Item Analysis and the Standard Error of Measurement, that are also significant to the study of language testing. They will be explained in future chapters. For an in-depth study of these statistical facts I recommend Brown (1996).

REFERENCES

Brown, J. D. (1996). *Testing in language programs*. Upper Saddle River, NJ: Prentice Hall Regents.

Hughes, A. (2003). *Testing for language teachers* (2nd ed.). Cambridge, England: Cambridge University Press.

HOW DO TEST SCORES AND DATA INTERPRETATION APPLY TO MY PROFESSION?

As a professional, it is your responsibility to *ensure* the accuracy, dependability, consistency, and stability of the scores that you obtain from any given assessment. Furthermore, it is your duty to ensure the appropriateness of the inferences, uses, trustworthiness, and legitimacy of claims of the results of any given assessment. Knowing how to interpret data and various scores is part of an educator's responsibility. It is impossible to summarize how your students or teachers are performing if you do not understand the basic concepts associated with data and what the data tell you.

Activity—Review standardized test scores you've recently obtained from your school, district, or state department of education. Identify the type of score(s), describe what the score(s) represent(s), how the score(s) apply/ies to you as a teacher or leader, and how you would explain the score(s) to an individual who knows nothing about test scores.

Figure Credit

Fig. 10.1: Center for Safe and Supportive Schools at Vanderbilt, from "Guidebook: School ClimateMeasurement, Data Use, and Improvement," p. 9. Center for Safe and Supportive Schools at Vanderbilt, 2014.

CHAPTER 11

Big Data, Data Visualization, and Data Illustration

What's the big deal with big data?

AS YOU READ

As you read the chapter, think about the following:

- How do **you** define or explain big data, visualization, and illustration?
- What is **your** experience with big data, visualization, and illustration?

WHAT ARE THE GENERAL PRINCIPLES OF BIG DATA, VISUALIZATION, AND ILLUSTRATION?

Big Data

Big data is a fairly new term that refers to data analytics. It stemmed from the 1980s–1990s educational movement of data-driven decision making (Picciano, 2012). Big data are collected via software tools that are often associated with assessment in education. However, assessment data are not the only forms of big data. If you think about students' use of everyday technology, you realize that students are "connected." They are using their phones, computers, social media, web searches, and others. Data are collected from all of these uses of technology, including educationally based computer software. There are multiple purposes and uses of these big data, some of which are perceived as positive and others as negative.

Data analytics inform companies, among other things, of usage, likes, dislikes, academic strengths, and academic weaknesses of each child. There have been great advancements in learning as a result of these data. Think about some of the software that your school uses to assess student knowledge in various disciplines. Does your school implement the use of software for students to read passages and then answer questions? The questions then change (get more difficult or less difficult) based on the individual responses to the questions? Does your school use online learning formats to teach students concepts? This is particularly prevalent in higher education with the use of various e-learning platforms. Do you utilize a flipped-classroom model? All of these are examples of learning environments that lead to the use of data analytics. As a result, educational institutions are exploring avenues of the use of big data technologies to manage data (Sin & Muthu, 2015).

Generally, big data refer to any data that are too large to manage with traditional applications that are not sophisticated enough to process the data (Sin & Muthu, 2015). Big data technologies use various tools and strategies to organize data so they can be interpreted. They may use various statistical analyses or simply organize data using various clusters. These organizational tools permit the audience to interpret the data to make appropriate decisions based on those data.

Pros and Cons of Big Data

There is controversy surrounding the use of big data. Think about the amount of data available and accessible about you as an individual. That may be scary to think about how much information companies and/or individuals can access about you. However, there may be some positive aspects of this as well.

If you think about the cons of big data, one of the first things that comes to mind may be the storage of such data. How secure are these data? What if there is a breach of the server that houses these data? What if the data are sold to others and the use of those data are different than the original purpose for collecting those data? These are all potential risks that are valid concerns. Thus, it is imperative that you are at least aware of these risks and that you do what you can to protect your students from such infringements. The cost of big data technologies may also be viewed as a con. Purchasing software, collecting, and managing these data can be expensive and the necessary resources are not always available or dedicating those resources to data may not be a priority for stakeholders when considering the overall budget for K–12 education.

Although there are potential risks, there are many uses of big data in education that may be viewed as pros. For example, big data provide opportunities for intelligent feedback. Intelligent feedback allows students to receive immediate feedback about their responses to computerized questions. The feedback may help improve student learning. Another common use and pro for education is the use of knowledge and/or skill acquisition. Various educational technology software systems are implemented for students to practice skills. The software then tracks students' responses and reports the depth of knowledge of the particular student. For a classroom teacher, this can be very useful information to help guide curriculum and instructional decisions. For educational leaders, you can use grade-level data to inform curriculum, instructional, evaluation, and professional development needs. Other examples of positive uses of big data include data visualization. The data analytics and illustration provided in various reports from these software companies can help educators interpret the data necessary to tell their story. There are several additional uses of big data that may be viewed as pros. These include identifying at-risk students, predicting performance of students or groups of students, grouping students, and identifying course selection for students (Sin & Muthu, 2015).

Data Visualization and Illustration

Data visualization can be thought of as a visual communication of a story. It helps your audience understand your story via data illustrations. There are multiple ways to illustrate data to make results easier to consume by your audience. To determine the best way to illustrate the data, you must first understand your audience, the point that is being made based on the data, the purpose for reporting the results, and so on. So, there cannot be a one-size-fits-all when discussing the best approaches to illustrating data. There are a few concepts that you must know to be able to make effective decisions about illustrating data.

Line Graphs

Are the data used to illustrate a trend over time? If the answer to this question is *yes*, then you will likely use a line graph to illustrate the data. Line graphs are primarily used to show multiple data collections over a period of time. You can have multiple lines on one graph or you can use multiple graphs. If you are trying to make comparisons over time, then it is typically best to use multiple lines on one graph.

Bar Charts

Are you trying to illustrate comparison data for a snapshot in time? If you answered *yes* to this question, then it is likely that you will use a bar graph to illustrate the data. Bar graphs are best used when you have data from one snapshot in time and illustrating differences is important. For example, if you have data from a battery of tests for a group, you may use a bar chart for each subtest to illustrate how the tests visually differ.

Histograms

Are you trying to illustrate the distribution of a set of scores? If you answered *yes* to this question, then you will likely illustrate the data by using a histogram. Histograms show the distribution of scores. A histogram will illustrate if the distribution is negatively or positively skewed and if the distribution is peaked or flat. Histograms are similar to bar charts but they group the data into ranges (frequencies).

Scatterplots

Are you trying to illustrate raw data points to determine if there is a linear relationship? If so, then it is likely that you will use a scatterplot.

Pie charts

Are you trying to illustrate categorical data? If you answered *yes* to this question, then you will likely use a pie chart. Pie charts are typically used when you have six or fewer categories of data that illustrate a proportion or a percentage of the whole.

WHAT DOES THE LITERATURE SAY ABOUT BIG DATA, VISUALIZATION, AND ILLUSTRATION?

Reading 11.1

LEARNING, SCHOOLING, AND DATA ANALYTICS

Ryan S. J. d. Baker

Since the 1960s, methods for extracting useful information from large data sets, termed *analytics* or *data mining*, have played a key role in fields such as physics and biology. In the last few years, the same trend has emerged in educational research and practice, an area termed *learning analytics* (LA; Ferguson, 2012) or *educational data mining* (EDM; Baker & Yacef, 2009). In brief, these two research areas seek to find ways to make beneficial use of the increasing amounts of data available about learners in order to better understand the processes of learning and the social and motivational factors surrounding learning. The goal of these efforts is to produce more efficient, more effective, and deeper learning in the context of increasingly positive learning experiences.

The emergence of EDM/LA is a recent phenomenon. The first meetings of scientists in this area were the Educational Data Mining workshops, which started in 2005 and became an annual conference series in 2008. This conference series was joined by the Learning Analytics and Knowledge conference series in 2011. The two research areas of EDM and LA, emerging from different communities of scientists and practitioners, have somewhat different goals; discussing these differences is outside the scope of this report (see Siemens & Baker, 2012). In brief, the validity of models of learners and learning is perhaps the key focus of the EDM community, whereas the use of the results of analysis to drive changes in practice by instructors is perhaps the key focus of the LA community. The conferences in EDM and LA were followed by the establishment of journals devoted to the topics, with the *Journal of Educational Data Mining* commencing publication in 2009 and the *International Journal of the Society for Learning Analytics Research* expected to commence publication in 2013. As of this writing, the International Educational Data Mining Society has approximately 150 members and over 600 subscribers on its mailing lists.

A range of methods has been developed by these two communities, drawing from areas such as data mining, computational science, statistics, psychometrics, and social network analysis. (A selection of these methods will be discussed below; a fuller review can be found in Baker & Siemens, in press).

RESEARCH SYNTHESIS

The methods of EDM have been applied to accomplish a range of objectives. This section reviews some of the applications which have had relatively large impacts or have relatively large potential, focusing on applications of particularly strong relevance to the readers of this *Handbook*.

One of the first applications of EDM was the development of models that could infer a student's knowledge as he or she worked through educational software. These inferences are in turn used to drive adaptation by the system. This application in fact preceded the existence of EDM or LA as research areas. Though student knowledge modeling began as a research area in the 1970s (Goldstein, 1979), the first model, which was both based on automated exploration of data and which achieved widespread dissemination in educational software, was Corbett and Anderson's (1995) Bayesian knowledge tracing (BKT) algorithm. One key difference between this algorithm and the types of student knowledge modeling used previously by the psychometrics community, for example in testing, was that BKT explicitly accounts for the fact that the student is learning at the same time he or she is being assessed; in other words, student knowledge is treated as a moving target. BKT was then incorporated into Cognitive Tutor software curricula for algebra and geometry (Koedinger & Corbett, 2006), sold by Carnegie Learning Inc., which was used by around 5% of U.S. high school students each year throughout the first decade of the 2000s. This software used BKT to decide when to advance the student on to new material, implementing an approach termed "mastery learning" (Bloom, 1968), in which the student does not advance until he or she demonstrates proficiency. By integrating BKT into mastery learning, Cognitive Tutor Algebra I was able to improve student test scores, with replication, in a range of settings (Koedinger & Corbett, 2006), although performance for geometry has been more mixed (Pane, McCaffrey, Slaughter, Steele, & Ikemoto, 2010). It is important to note that the automated algorithms and learning support in Cognitive Tutor replaced the workbook rather than the teacher; in Cognitive Tutor classrooms, the teacher spends more time interacting with students in one-on-one learning support sessions than in full-class teaching (Schofield, 1995), perhaps another reason for this approach's success.

Since the implementation of Cognitive Tutors, new online learning systems have added emphasis on providing actionable and formative information to teachers. For example, the ASSISTment system (it "assesses while it assists") has created a reporting system that teachers can use to determine both what material specific students are struggling with and what items the entire class is struggling with (Feng & Heffernan, 2006). Teachers using the system review student homework before class and are able to change the focus of classroom activities based on data on student understanding, leading to better classroom performance than is seen with traditional homework (Koedinger, McLaughlin, & Heffernan, 2010; Mendicino, Razzaq, & Heffernan, 2009).

> *Teachers using the system review student homework before class and are able to change the focus of classroom activities based on student understanding ...*

The types of formative information that can be assessed by online learning systems have gone beyond just student knowledge in recent years. Algorithms for assessing disengaged behaviors have been developed for learning systems recently (Baker, 2007; Baker, Corbett, & Koedinger, 2004; Pardos, Baker, San Pedro, Gowda, & Gowda, 2013; San Pedro, Baker, & Rodrigo, 2011), making it possible to assess with reasonable accuracy whether students are careless, off-task, or intentionally misusing educational software, among other disengaged behaviors. These algorithms have been extended to also infer student emotion during learning, just from data readily available to computer systems (i.e., no physiological sensors; see Baker et al., 2012; Pardos et al., 2013; Sabourin, Rowe, Mott, & Lester, 2011). As these models are built into systems such as ASSISTments or Crystal Island, increasing amounts of information will be available to classroom teachers; the key challenge will be providing it to teachers in useful and timely fashions.

Another direction for integrating EDM and LA research into educational practice is to predict student dropout and course failure, a step towards providing early intervention. One particularly successful example is the Purdue Signals Project, reported to have significantly improved student outcomes at Purdue University (Arnold, 2010). This system uses prediction models to infer early in the semester which students are likely to fail or drop out of a course; a list of students at risk is generated and sent to an instructor, along with recommended template emails for these students which inform them about help resources available. This type of system is being implemented at an increasing number of universities, both in independent projects (Ming & Ming, 2012) and through a commercial vendor, Ellucian, which is distributing the Signals software to additional universities.

While dropout and failure prediction at the K–12 level have not yet reached the level of deployment and demonstrated success of the Purdue Signals Project, there are several examples of successful prediction of student dropout at the K–12 level. To give just a few examples, Tobin and Sugai (1999) predict high school dropout from middle school disciplinary records; Bowers (2010) uses changes in student achievement to predict high school dropout as early as third grade; San Pedro, Baker, Bowers, and Heffernan (in press) use data on middle school student emotion and learning within the aforementioned ASSISTment system to predict which students will attend college. Each of these approaches has the potential to be used at scale; the challenges to doing so are organizational rather than technical.

Beyond supporting specific changes in practice, EDM and LA research has played an increasingly important role in supporting basic discovery in education research. The opportunity to leverage very fine-grained data (often multiple data points per minute) across entire years of data for a specific student, in combination with automated methods for sifting through that data, has been an excellent opportunity for better understanding learners and learning. Types of EDM methods, such as discovery with models and structure discovery algorithms, have enabled a variety of analyses, including discovery of which exploratory learning strategies are most effective (Amershi & Conati, 2009), which patterns of group work lead to more successful group projects (Perera, Kay, Koprinska, Yacef, & Zaiane, 2009), which meta-cognitive behaviors lead to deep learning (Baker et al., 2012), and how small-scale choices in the design of educational software can lead to substantial differences in student engagement (Baker et al., 2009).

The opportunity to leverage very fine-grained data ... across entire years of data for a specific student ... has been an excellent opportunity for better understanding learners and learning.

ACTION PRINCIPLES

In this section, I propose a set of action principles for schools, local education agencies (LEAs), and state education agencies (SEAs), suggesting how the emerging fields of learning analytics and educational data mining can be used to improve their practice.

Action Principles for Schools

Provide Formative Data to Teachers on Student Learning

In recent years, the advent of learning systems such as ASSISTments (but also Cognitive Tutors, Reasoning Mind, Aleks, LearnBop, and many others) has presented an opportunity to provide teachers with considerably more information on their students' learning, generally in easy-to-interpret formats. Depending on a school's goals, some of these systems (such as Cognitive Tutors and Reasoning Mind) can be adopted as an entire curriculum; others, such as ASSISTments and LearnBop, simply replace existing homework or seatwork and can be used with a variety of curricula.

These systems provide teachers with information on which students are struggling and what they are struggling on. This enables teachers to identify what material these students need support with, so that the teacher can provide them with extra assistance (Schofield, 1995). Sometimes, a teacher can also see by using these systems that a specific topic is difficult for all students; this is also possible to determine when the teacher grades by hand, but the teacher is informed earlier by automated systems, thus supporting timely intervention.

Predict Which Students Are at Risk for Dropping Out

As discussed above, one of the key successes of learning analytics at the undergraduate level is predicting which students are at risk of failing or dropping out. At that level, success has been achieved not only in predicting who is at risk but also in embedding this information in effective interventions used to reduce dropout (Arnold, 2010).

Several research projects have demonstrated that the same type of prediction is possible for K–12 schools. The work by Bowers (2010) in predicting high school dropout from grades students receive in elementary school demonstrates that this type of prediction is possible just from the data already available in schools. Similarly, data on disciplinary referrals (e.g., fighting) during middle school can predict who will drop out in high school (Tobin & Sugai, 1999). However, both of these types of indicators may be identifying students at very high risk, students whose problems are outside those that are easily addressed by schools. Dropout prediction from interactions with educational software may provide a way to identify at-risk students whose challenges can be more easily addressed, and may provide more precise information on the factors causing those students to be at risk. For example, recent work has indicated that educational software can infer not just student knowledge but also multiple dimensions of student engagement. Long-term prediction from educational software is still emerging

(San Pedro et al., in press) but is likely to be available in an increasing number of educational software packages used in schools in the years to come.

Identify Learning Topics that Are Being Learned Less Well Within School

Recent educational software is able to identify which skills and topics are being learned less well than others within a specific classroom or school. This type of information is available in reports from many modern learning software packages, including but not limited to ASSISTments, the Cognitive Tutor, LearnBop, and Reasoning Mind. This type of information does not require using a software package—it is possible to think of teachers across schools recording homework data, tagging it by topic, and looking together for topics where performance is poor—but it is much easier to do in schools and classrooms that use educational software since the bookkeeping and data integration is offloaded to a computer system.

Understanding the topics for which a school's current curriculum and pedagogical approaches are working less effectively creates opportunities to redesign teaching in those areas or to supplement current practice with other resources. If a school is generally performing poorly on division of fractions across teachers, for example, it is probably not a flaw in one person's teaching but instead a flaw in the curriculum being used, a flaw that can be addressed throughout the school.

Capture and Respond to Changes in Student Engagement

In 2013, the automated assessment of student engagement and emotion remains primarily within research classrooms, but it is emerging within a range of learning systems, making it likely that it will become generally available in classrooms in the coming years. As automated assessment of student engagement and emotion becomes increasingly feasible to integrate within online learning systems, such as ASSISTments, it is likely to become useful to teachers. When it is available, teachers and school psychologists will be able to use it to identify early students who have become disengaged across classes, potentially identifying a student in need of an intervention. If problem behaviors are below the threshold of office referrals, a student's general changes in behavior may not be noticed; with this type of technology, it may be possible to identify shifts in engagement quickly. Even within a single class, emotion- and engagement-sensing technology may prove quite useful. For example, if a teacher can identify that a student was frustrated during his or her online homework the night before, it may be possible to talk to the student to better understand why the material was particularly difficult.

Action Principles for Local Education Agencies

Identify Specific Areas of Excellence and High Success in Teaching Practice

When educational software that assesses engagement and learning is used in schools, it can be beneficial not just to individual teachers and schools but to local education agencies (LEAs) as well. This type of assessment can provide information that can help LEAs to identify teachers that are successful in promoting engagement and learning in specific areas. The expertise these teachers have can then be leveraged by their LEA. For example, if a teacher is succeeding at teaching a topic that other teachers are known to struggle with—as manifested by better performance by his/her students on that topic—that teacher could give a brief workshop on his or her teaching strategies. Similarly, if one teacher's classes generally experience less boredom (while learning equally well), it may be worth having this teacher mentor other teachers in engaging their students. In this fashion, it may be possible to identify exact areas of excellence and share them across a school district.

Although these methods can be used to identify exemplary teachers, rewarding teachers who are particularly successful according to these types of internal measures may have undesired effects. If teacher pay were linked to evidence of frustration in a system like ASSISTments, some teachers might alter their classroom practice in undesirable ways, to try to "game the system," for example, by walking around the classroom, immediately giving answers to every struggling student. Even if this did not improve the assessment of engagement by the software, it might still be attempted, with unpredictable and likely undesirable results. Automated detectors will be more effective, and more useful, if there is not an attempt to subvert them (necessitating automated detectors of subversion, as seen in Baker et al., 2004). In sum, integrating automated assessment systems into reward structures has the potential to reduce their effectiveness for other goals.

Identify Students Who Could Benefit from Enrichment Programs

Another upcoming opportunity for LEAs is to identify specific students who could benefit from enrichment programs. Across the U.S., after-school, weekend, and summer programs are available to learners, funded by federal agencies—such as the National Science Foundation's Innovative Technology Experiences for Students and Teachers (ITEST) program—state agencies, foundations, and private funders. However, there remains insufficient capacity to provide enrichment programs to all students who want to enroll in them, and the students who do enroll are often drawn from wealthier groups (Gardner, Roth, & Brooks-Gunn,

2009). In addition, not all enrichment programs are the same; there is a question of fit when selecting students for an enrichment program.

When technology becomes readily available to assess engagement in class, it will be increasingly possible to identify students who are highly engaged in specific subjects. These students—especially if they are disadvantaged—should be particularly strong candidates for enrichment programs, and efforts should be made to place them in enrichment programs that fit their interests and will help them develop their interests in these specific areas.

Develop Internal Expertise in Learning Analytics

A third recommendation for local education agencies is to develop internal expertise in learning analytics. In recent years, there has been an explosion of data that can be used for a wide range of purposes, as indicated in the recommendations above (both those for schools and for local education agencies). Local education agencies can play an essential role in fulfilling both of these recommendations, conducting analyses at the district level and supporting schools in conducting school-level analyses (or even conducting analyses for schools).

The cost of hiring one or more learning analytics experts or of training an existing member of the LEA in learning analytics methods may in the future be seen as a relatively small expense in relation to the benefits that can be achieved. There are increasing opportunities to train LEA personnel, including the upcoming fall 2013 massive online open course (MOOC) within Coursera, Big Data in Education, and an annual MOOC on learning analytics provided by the Society for Learning Analytics Research. Also, an increasing number of graduate programs specialize in this area. As of this writing, programs in learning analytics or related areas are offered at Teachers College Columbia University, Carnegie Mellon University, and Worcester Polytechnic Institute, programs creating an increasing pool of trained individuals who can provide this type of expertise to schools.

Develop Data Management and Sharing Plans to Support Partnerships With University Researchers in Line With Legal Obligations

Beyond hiring their own staff in learning analytics, school districts may be able to leverage the expertise of universities. There is growing pool of university faculty, postdoctoral researchers, and graduate students who are deeply interested in learning analytics and EDM and want to use these methods to benefit American education, at a wide range of institutions, even beyond those officially offering training in these areas. These researchers are a resource that LEAs can leverage to conduct analyses beyond their own capacities. Such collaborations are likely to benefit all but the largest and wealthiest school districts; even for those districts, there may be expertise in learning analytics located in specific university research groups that is duplicated nowhere else.

However, these collaborations will not occur unless appropriate institutional, legal, and infrastructural arrangements are made. One key step is the creation of procedures for quickly de-identifying data sets (removing all potentially identifying information) so they can be shared with university researchers without violation of relevant federal privacy laws and guidelines, such as the Family Educational Rights and Privacy Act (FERPA), the federal law that protects the privacy of student education records. Creating procedures for sending de-identified data to researchers but being able to link findings from those researchers back to individual students within the LEA will be essential in order to benefit those students, using the information obtained in research. Policies for such de-identification would prevent identifiable information from being transmitted outside the school district and designate some individual within the LEA to hold a strictly guarded key, so that the findings can be tracked back to students within the LEA. In addition, LEAs should instruct their institutional review boards to follow relevant federal law and guidelines for fast-tracking research with minimal risk of harm to students, for research projects classified as exempt from review or fit for expedited review under the federal guidelines. Currently, many LEAs—particularly in larger cities—choose not to follow federal guidelines for review of research, instead creating onerous review processes that lead many research groups to avoid working with those LEAs. The result is that students in suburban school districts benefit more from the university researchers in major urban centers than students in those urban centers, reinforcing inequities. Even after approving research, many LEAs currently require extensive legal agreements, again well beyond federal or state requirements, delaying or preventing research collaborations. In general, streamlining procedures for learning analytics research (while following all federal laws and guidelines, and protecting student privacy) is likely to benefit students considerably and facilitate the task of LEAs in supporting their students.

Action Principles for State Education Agencies

Capture Data About Students According to Broad-Based Range of Indicators

One important way that state education agencies (SEAs) can support learning analytics is by taking steps to collect a broad range of types of data. Many types of data are now available about learners and schools beyond what routinely make it to state education agencies—from log files, to automated assessments, to data from classroom observations. By having a range of types of data, SEAs will be able to conduct analyses of the factors leading to better performance on state standardized exams, higher college attendance, and so on. States should partner with resource centers to select which indicators to capture and encourage vendors to provide understandable and reasonably complete data to their SEA as a condition. Similarly, SEAs should incentivize schools and LEAs to also collect a broad range of data and provide it to SEAs. An SEA's data is unlikely to reach its full potential except in partnership with LEAs that are collecting a broad range of useful data.

While different schools may collect and use data that is not fully compatible, making sure that all of this data is available at the state level will be a useful step towards supporting state-level analyses. For example, even if one learning system tends to predict higher engagement than another learning system, having data from both learning systems will make it possible to see statewide trends.

Form Practices for Aligning Student Data Even in the Face of Mobility

School mobility is a fact of 21st-century education; because American society is highly mobile, students are likely to change schools repeatedly during their education. While school mobility may not be problematic for students of high socioeconomic status (SES), it is associated with poorer outcomes among lower SES and minority students, especially if a student changes schools several times (Xu, Hannaway, & D'Souza, 2009). Mobility can also be a problem for tracking students and applying learning analytics to the data from these students; it is easier to obtain data for and therefore apply predictive models to students who do not change school districts, implying that prediction of at-risk status will be least effective for students who are already at risk due to their mobility.

State education agencies can play a key role in tracking these students by using state-level identifiers to track student progress even if the student moves. Equally importantly, SEAs should encourage LEAs and schools to store all data in terms of state-level identifiers and should support LEAs and schools in obtaining student data from other LEAs and schools (ideally through state-level databases that all LEAs provide data to and draw from). In that fashion, learning analytics analyses can leverage all of the data available for a specific student.

States can further support local districts by identifying effective practices for forming partnerships with university and corporate researchers focused on data use. As discussed above, several benefits may accrue to LEAs in forming partnerships with university researchers. SEAs have a key role to play in setting the tone for collaboration and nudging LEAs to develop and conduct these partnerships appropriately. SEAs should educate LEAs about—and encourage them to follow—federal and state guidelines so that LEAs avoid unnecessary and unproductive roadblocks which prevent interventions that would benefit students, while also avoiding violating federal or state laws or violating student privacy.

Identify Exemplary Teachers and Schools

SEAs, even more than LEAs, have the potential to identify schools or teachers who are succeeding at promoting engagement and learning in specific areas. Across a state, there are likely to be exemplary practices, often in unexpected places, that can be identified through learning analytics. These practices can then be studied and communicated across the state in collaboration with resource centers. It is worth noting that, as with LEAs, the indicators that are useful for these types of analyses are better used in a formative fashion than to drive financial incentives (or firings); the incentives for gaming the system or even cheating are substantial if financial incentives are used and doing so would reduce the potential for disseminating exemplary practices statewide.

Identify Regional Gaps in Enrichment Programs

As discussed above, enrichment programs are not currently available to all students who want to enroll in them, and the students who do enroll are often drawn from wealthier groups (Gardner, Roth, & Brooks-Gunn, 2009), in part due to regional disparities. While some of the factors leading to these differences are difficult to address (e.g., parental choice and funding choices made by private foundations and individuals), better data on where the needs are may help to influence the allocation of government resources and potential private funding as well. By identifying the number of at-risk students and students likely to benefit from programs, and comparing these numbers to the availability of program slots in different regions, SEAs will be able to identify which regions have an insufficient quantity of enrichment programs and support program expansion and creation. Simply publishing data on where needs exist is likely to influence funding decisions, not just by private foundations and individuals but by programs funded by the federal government. For example, federal programs like National Science Foundation's ITEST might be more likely to fund programs in specific regions declared in need by SEAs than in regions shown to have a relative oversupply of enrichment programs.

Learning analytics may also have the potential to identify more quickly which enrichment programs are working. If an enrichment program is provided to elementary school students, any evidence of its effect on high school dropout rates or college attendance is a distant prospect. Obtaining data on learning and engagement from schools in the year following a student's participation in an enrichment program may provide more rapid signals as to which programs are succeeding in their goals.

ACKNOWLEDGEMENTS

The author would like to thank Lisa Rossi, Janet Twyman, Marilyn Murphy, and Stephen Page for helpful comments and suggestions.

REFERENCES

Amershi, S., & Conati, C. (2009). Combining unsupervised and supervised machine learning to build user models for exploratory learning environments. *Journal of Educational Data Mining, 1*(1), 71–81.

Arnold, K. E. (2010). Signals: Applying academic analytics. *Educause Quarterly, 33*, 1–10.

Baker, R. S. J. d. (2007). *Modeling and understanding students' off-task behavior in intelligent tutoring systems.* In Proceedings of ACM CHI 2007: Computer-human interaction, 1059–1068.

Baker, R. S., Corbett, A. T., & Koedinger, K. R. (2004). *Detecting student misuse of intelligent tutoring systems.* Proceedings of the 7th International Conference on Intelligent Tutoring Systems, 531–540.

Baker, R. S. J. d., de Carvalho, A. M. J. A., Raspat, J., Aleven, V., Corbett, A. T., & Koedinger, K. R. (2009). *Educational software features that encourage and discourage "gaming the system."* Proceedings of the 14th International Conference on Artificial Intelligence in Education, 475–482.

Baker, R. S. J. d., Gowda, S. M., Wixon, M., Kalka, J., Wagner, A. Z., Salvi, A., ... Rossi, L. (2012). *Sensor-free automated detection of affect in a Cognitive Tutor for algebra.* Proceedings of the 5th International Conference on Educational Data Mining, 126–133.

Baker, R. S. J. d., & Siemens, G. (in press). Educational data mining and learning analytics. In K. Sawyer (Ed.), *Cambridge handbook of the learning sciences* (2nd ed.). Cambridge, MA: Cambridge University Press.

Baker, R. S. J. d., & Yacef, K. (2009). The state of educational data mining in 2009: A review and future visions. *Journal of Educational Data Mining, 1*(1), 3–17.

Bloom, B. S. (1968). Learning for mastery. *Evaluation Comment, 1*(2), 1–12.

Bowers, A. J. (2010). Analyzing the longitudinal K-12 grading histories of entire cohorts of students: Grades, data driven decision making, dropping out, and hierarchical cluster analysis. *Practical Assessment, Research & Evaluation (PARE), 15*(7), 1–18.

Corbett, A. T., & Anderson, J. R. (1995). Knowledge tracing: Modeling the acquisition of procedural knowledge. *User Modeling and User-Adapted Interaction, 4*, 253–278.

Feng, M., & Heffernan, N. T. (2006). Informing teachers live about student learning: Reporting in the ASSISTment system. *Technology, Instruction, Cognition, and Learning Journal, 3*(1–2), 1–8.

Ferguson, R. (2012). *The state of learning analytics in 2012: A review and future challenges* (Tech. Rep. No. KMI-12-01). Milton Keynes, UK: Open University, Knowledge Media Institute. Retrieved from http://kmi.open.ac.uk/publications/techreport/kmi-12-01

Gardner, M., Roth, J. L., & Brooks-Gunn, J. (2009, October). *Can after-school programs help level the playing field for disadvantaged youth?* (Equity Matters: Research Review, 4). New York, NY: Campaign for Educational Equity, Teachers College, Columbia University.

Goldstein, I. J. (1979). The genetic graph: a representation for the evolution of procedural knowledge. *International Journal of Man-Machine Studies, 11*(1), 51–77.

Koedinger, K. R., & Corbett, A. T. (2006). Cognitive tutors: Technology bringing learning science to the classroom. In K. Sawyer (Ed.), *The Cambridge handbook of the learning sciences* (pp. 61–78). New York: Cambridge University Press.

Koedinger, K. R., McLaughlin, E. A., & Heffernan, N. T. (2010). A quasi-experimental evaluation of an on-line formative assessment and tutoring system. *Journal of Educational Computing Research, 43*(4), 489–510.

Mendicino, M., Razzaq, L., & Heffernan, N. T. (2009). A comparison of traditional homework to computer-supported homework. *Journal of Research on Technology in Education, 41*, 331–358.

Ming, N. C., & Ming, V. L. (2012). *Predicting student outcomes from unstructured data.* Proceedings of the 2nd International Workshop on Personalization Approaches in Learning Environments, 11–16.

Pane, J. F., McCaffrey, D. F., Slaughter, M. E., Steele, J. L., & Ikemoto, G. S. (2010). An experiment to evaluate the efficacy of Cognitive Tutor Geometry. *Journal of Research on Educational Effectiveness, 3*(3), 254–281.

Pardos, Z. A., Baker, R. S. J. d., San Pedro, M. O. C. Z., Gowda, S. M., & Gowda, S. M. (2013). *Affective states and state tests: Investigating how affect throughout the school year predicts end of year learning outcomes.* To appear in Proceedings of the 3rd International Conference on Learning Analytics and Knowledge.

Perera, D., Kay, J., Koprinska, I., Yacef, K., & Zaiane, O. R. (2009). Clustering and sequential pattern mining of online collaborative learning data. *IEEE Transactions on Knowledge and Data Engineering, 21*(6), 759–772.

Sabourin, J., Rowe, J., Mott, B., & Lester, J. (2011). *When off-task in on-task: The affective role of off-task behavior in narrative-centered learning environments.* Proceedings of the 15th International Conference on Artificial Intelligence in Education, 534–536.

San Pedro, M. O. C., Baker, R., & Rodrigo, M. M. (2011). *Detecting carelessness through contextual estimation of slip probabilities among students using an intelligent tutor for mathematics.* Proceedings of 15th International Conference on Artificial Intelligence in Education, 304–311.

San Pedro, M. O. Z., Baker, R. S. J. d., Bowers, A. J., & Heffernan, N. T. (in press). *Predicting college enrollment from student interaction with an intelligent tutoring system*

in middle school. To appear in Proceedings of the 6th International Conference on Educataional Data Mining.
Schofield, J. W. (1995). *Computers and classroom culture*. Cambridge, MA: Cambridge University Press.
Siemens, G., & Baker, R. S. J. d. (2012). Learning analytics and educational data mining: Towards communication and collaboration. *Proceedings of the 2nd International Conference on Learning Analytics and Knowledge*, 252–254.
Tobin, T. J., & Sugai, G. M. (1999). Using sixth-grade school records to predict violence, chronic discipline problems, and high school outcomes. *Journal of Emotional and Behavioral Disorders, 7*(1), 40–53.
Xu, Z., Hannaway, J., & D'Souza, S. (2009). *Student transience in North Carolina: The effect of school mobility on student outcomes using longitudinal data* (CALDER Working Paper 22). Washington, DC: The Urban Institute.

Reading 11.2

PIVOT TABLE APPLICATIONS

Pivot Tables—Applications in Descriptive Statistics and Data Analysis

Amar Sahray

The ***pivot table*** is one of the most powerful and easy-to-use tools in EXCEL. This application is particularly helpful in extracting large amount of useful information from a data set. Using EXCEL's pivot table you can:

- Create cross-tabulation or break the data down by categories so that you can see the customer rating by gender, or see the total dollars spent on online orders by the day of the week, region of the country, time of the day, or a combination of these.
- Create tables of counts to display the sale of a product by brand name, or number of customers by region. You can create cross-tabulation or contingency tables to display the counts only. The pivot tables provide more flexibility compared to cross tables. The pivot tables can be used to display counts, sums, averages, and other summary measures of the data.

APPLICATION 1: USE PIVOT CHART TO SUMMARIZE QUANTITATIVE DATA—CREATING A FREQUENCY DISTRIBUTION AND A HISTOGRAM

Earlier, we explained how to construct a frequency distribution and a histogram using the **Data** → **Data Analysis** → **Histogram** option. Using the Excel's pivot chart, we can develop both a frequency distribution and a histogram at the same time. The steps are outlined in Table 11.1.

TABLE 11.1 **Instructions for Creating a Frequency Distribution and Histogram using Pivot Chart**

1. Open the worksheet **PivotTable1.xlsx**. This file contains lifetime or the survival time of a sample of 200 computer chips from a laboratory test.
2. Click the **Insert** tab on the EXCEL main menu.
3. In the **Charts** group, click the downward pointing arrow and select **Pivot Chart & Pivot Table**.
4. **Create PivotTable** dialog box will be displayed. In this dialog box: Click on the circle to the left of **Select a table or range** Click on the **Table/Range** box and Select the range by clicking on cell A1 and dragging down the cursor to the last value, which is A201
5. Click on the circle to the left of **Existing Worksheet** (to display the **Pivot Table** and **PivotChart** on the existing work sheet) Click in the **Location** box and then click on cell **C4** (or any blank cell) of the worksheet Click **OK** You will see boxes entitled **Pivot Table 1, Chart 1,** and **Pivot Chart Fields** displayed
6. In the **PivotTable Field List** go to **Choose Fields to add to report** Click on **Lifetime (Hours)** and drag this field to the **AXIS (Categories) box** Next, drag the **Lifetime (Hours)** field to the **Values box** (Note that a field is simply a variable name or a column of data)
7. Click the downward pointing arrow to the right of **Sum of Lifetime (hours)** in the **Values** box.
8. Click **Value Field Settings** from the list of options.
9. The **Value Field Settings** dialog box will appear under **Summarize value field by**, select **Count** Click **OK**
10. Right click cell C5 in the pivot table or any other cell containing the **Life Time** (Hours) values.
11. Select **Group** from the list of options.
12. The **Grouping** dialog box will be displayed. In this box Enter **200** in the **Starting at** box Enter **440** in the **Ending** at box Enter **30** in the By box Click **OK** (a pivot chart and a table will be displayed) (Note the starting value of 200 and the ending value of 440 with a width of 30 are selected so that the minimum and maximum value in the data are contained in this range. You need to select these values by sorting the data and deciding on the number of classes and the width of the interval.)
13. When the chart is displayed, click anywhere in the chart then click the **Design** tab on the main Menu and select a design (we selected the last one).

TABLE 11.1 **Instructions for Creating a Frequency Distribution and Histogram using Pivot Chart (*Continued*)**

14.	Double click on any bar of the histogram to display **Format Data Series**. Drag the **Gap Width** button to the extreme left or type 0 in the Gap Width box to close the gaps in the histogram.
15.	From the top right of the displayed histogram, click the box with **+** sign. If the box is not displayed, click on the chart and the tab will appear.
16.	Check the **Axes**, **Axis Title**, **Chart Title**, and **Data Labels**. Type the axes titles and the chart Title shown in Figure 2A.15. Selecting **Data Labels** will put the values of each category on the top of the bar.
17.	Right click on any bar of the displayed chart. Select **Fill** to change the color and the type of fill of the bars.
18.	Repeat step 17 and select **Outline** to change the outline or the color or the outline of the bars of the histogram.
19.	Two buttons—**Count of Lifetime (Hours)** and **Lifetime (Hours)**—are displayed on the graph. To hide these, right click on any button and select **Hide all Field Buttons on Chart**.
20.	Double click on any *x*-axis value. A **Format Axis** dialog box will appear with several options. In the **Custom** box type **−50** to change the orientation of the *x*-axis values. Your frequency table and the histogram will be as shown in Table 11.2 and Figure 11.1.

TABLE 11.2

ROW LABELS	COUNT OF LIFETIME (HOURS)
200–229	4
230–259	11
260–289	30
290–319	46
320–349	49
350–379	40
380–409	14
410–440	6
Grand Total	200

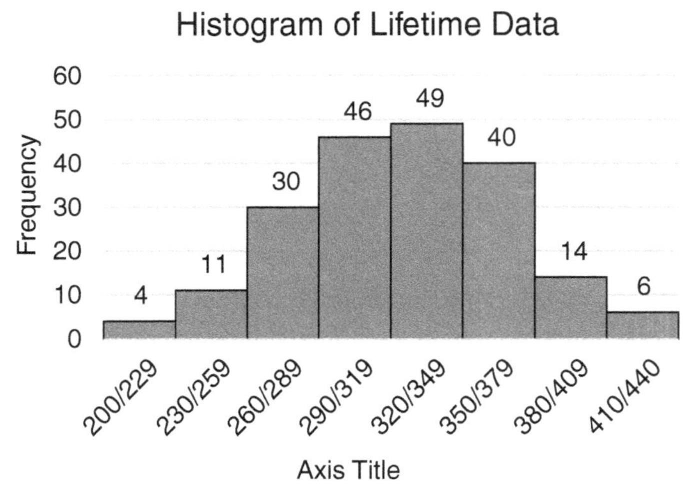

FIGURE 11.1 **Histogram of Lifetime data**

APPLICATION 2: USE A PIVOT CHART TO SUMMARIZE QUALITATIVE OR CATEGORICAL DATA

Develop a frequency distribution and graphical display (a bar chart and a pie chart) for the categorical data in the data file **PivotTable3.xlsx**. This data file contains the name brand television sets sold by an electronic retailer over a period of 2 months. The management would like to know how many televisions of each brand were sold. Follow the steps in Table 11.3.

TABLE 11.3

1. Open the worksheet **PivotTable3.xlsx**. This file contains different brands of televisions sold over a period of 2 months.
2. Click the **Insert** tab on the EXCEL main menu.
3. In the **Charts** group, click the downward pointing arrow and select **Pivot Chart & Pivot Table.**
4. **Create PivotTable** dialog box will be displayed. In this dialog box

   Click on the circle to the left of **Select a table or range**

   Click on the **Table/Range** box and select the range by clicking on cell A1 and dragging down the cursor to the last value which is A151
5. Click on the circle to the left of **Existing Worksheet** (to display the **Pivot Table** and **PivotChart** on the existing work sheet).

Click in the **Location** box, then click on cell **C4** (or any blank cell) of the worksheet Click **OK**. You will see boxes entitled **Pivot Table 3 and Chart** displayed. On the right pane **Pivot Table Fields** will be displayed (if you don't see **Pivot Table Fields** on the right side, click on **Pivot Table** box)

6. In the **PivotTable Field List** go to **Choose Fields to add to report**

   Click on **TV Brands Sold** and drag this field to the **ROWS box**

   Next, drag the **TV Brands sold** field to the **VALUES box**

   The pivot table and the bar chart will be displayed.

7. Click on the displayed bar chart, select **Design** from the main menu and choose the desired design from the options.
8. From the top right of the displayed bar chart, click the box with + sign. If the box is not displayed, click on the chart and the tab will appear.
9. Check the **Axes**, **Axis Title**, **Chart Title**, and **Data Labels**. Type the axes titles and the chart title shown in Figure 11.2. Selecting **Data Labels** will put the values of each category on the top of the bar (some designs will automatically print the values on the top of each bar).
10. Right click on any bar of the displayed chart. Select **Fill** to change the color and the type of fill of the bars.
11. Repeat step 10 and select **Outline** to change the outline or the color of the outline of the bars of the histogram.
12. **Create a pie chart of the data:** To create a pie chart select the two columns of the pivot table (the brands under **Row Labels** and **Count of TV Brands Sold**)
13. From the main menu, click on **Insert** and from the charts group click on **Pie** and select the first option under **2-D Pie**
14. A **Pie chart** of the pivot table data will be displayed. Click on the pie chart then select **Design** from the main menu and select the design of your choice.
15. From the top right of the displayed pie chart, click the box with + sign. If the box is not displayed, click on the chart and the tab will appear.
16. Edit your histogram by selecting appropriate boxes. On the top left of the chart a button—**Count of TV Brand Sold** is displayed. To hide this, right click on it and select **Hide all Field Buttons on Chart**.

Table 11.4 shows the pivot table. The bar chart and the pie chart are shown in Figures 11.2 and 11.3. Note that your charts may look different depending on the designs you selected.

TABLE 11.4

ROW LABELS	COUNT OF TV BRANDS SOLD
Mitsubishi	18
Panasonic	45
RCA	8
Sanyo	13
Sharp	19
Sony	47
Grand Total	150

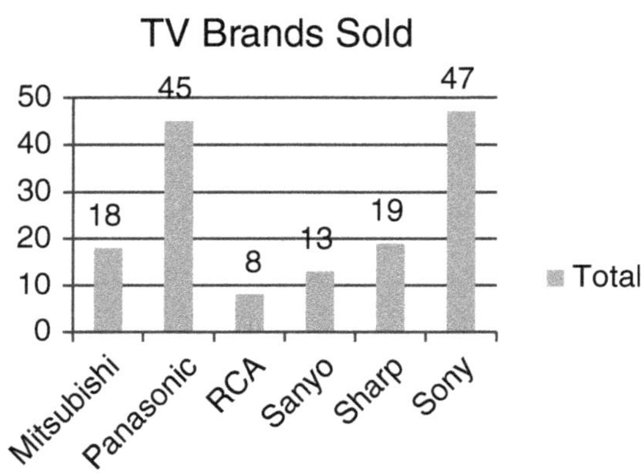

FIGURE 11.2

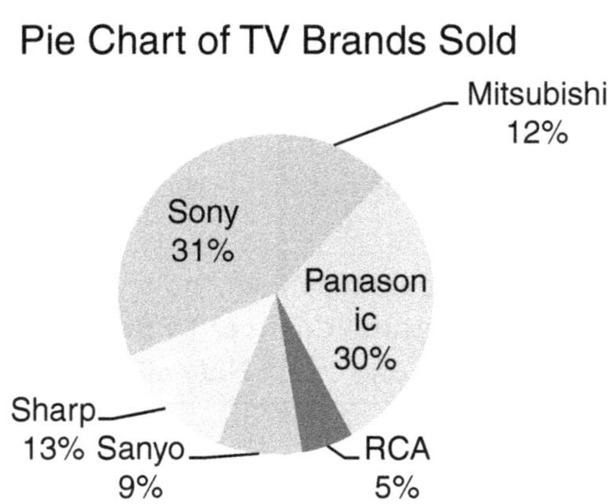

FIGURE 11.3

APPLICATION 3: CROSS-TABULATION WITH TWO AND THREE QUALITATIVE OR CATEGORICAL DATA USING PIVOT TABLE OPTION

The pivot table can also be used to summarize the data for two or more variables simultaneously. This exercise will enable you develop a cross-tabulation of the data provided in the data file **PivotTable2.xlsx**. This file contains data on student major by gender and employment status.

Develop a cross-tabulation to summarize the data by gender and student major, that is, create a pivot table that will have gender as the row variable and student major as the column variable. Create a chart of your table. Follow the steps in Table 11.5 to create such a table.

TABLE 11.5

1. Open the worksheet **PivotTable2.xlsx**.
2. Click the **Insert** tab on the EXCEL main menu.
3. In the **Charts** group, click the downward pointing arrow and select **Pivot Chart & Pivot Table**.
4. **Create PivotTable** dialog box will be displayed. In this dialog box

   Click on the circle to the left of **Select a table or range**

   Click on the **Table/Range** box and select the range by clicking on cell A1 and dragging down the cursor to the last value, which is C201
5. Click on the circle to the left of **Existing Worksheet** (to display the **Pivot Table** and **PivotChart** on the existing work sheet).

   Click in the **Location** box then click on cell **E5** (or any blank cell) of the worksheet Click **OK**

   You will see boxes entitled **Pivot Table # and Chart #** displayed. On the right pane **PivotTable Fields** will be displayed (if you don't see **Pivot Table Fields** on the right side, click on **Pivot Table** box)
6. In the **PivotTable Field List** go to **Choose Fields to add to report**

   Click on **Gender** and drag this field to the **ROWS box**

   Next, drag the **Major** field to the **COLUMNS box**

   Next, drag the **Major** to the **VALUES box**

   In the Values box, click on the downward pointing arrow and select **Value Field Setting**
7. In the **Value Field Setting** box, select **Count**

   The pivot table of student major and gender along with the bar chart will be displayed.

8. Click on the displayed bar chart, select **Design** from the main menu, and choose the desired design (we selected the last one).
9. From the top right of the displayed bar chart, click the box with + sign. If the box is not displayed, click on the chart and the tab will appear.
10. Check the **Axes**, **Axis Title**, **Chart Title**, and **Data Labels**. Type the axes titles and the chart title if needed. Selecting **Data Labels** will put the values of each category on the top of the bar (some designs will automatically print the values on the top of the bars).

Note: On the top left of the chart three buttons—**Count of Major, Major and Gender**—are displayed.
To hide these, right click on any button and select **Hide all Field Buttons on Chart**.

Table 11.6 shows the pivot table. The bar chart is displayed in Figure 11.4.

TABLE 11.6

COUNT OF MAJOR	COLUMN LABELS					
ROW LABELS	1	2	3	4	5	GRAND TOTAL
Female	6	11	34	32	9	102
Male	12	28	11	31	16	98
Grand Total	18	39	45	63	35	200

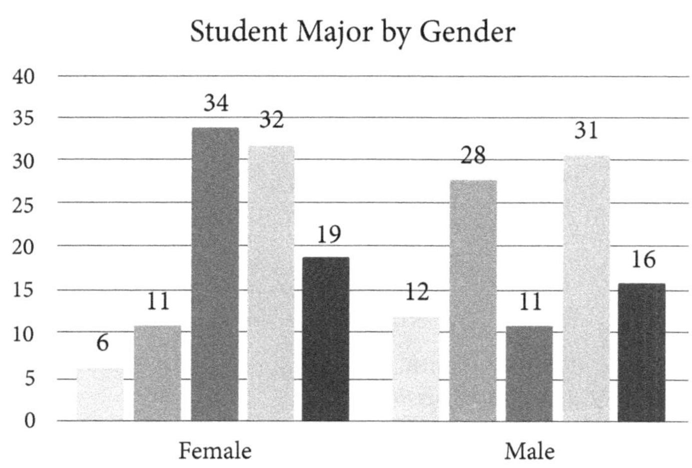

FIGURE 11.4 **Student Major by Gender**

HOW DO BIG DATA, VISUALIZATION, AND ILLUSTRATION APPLY TO MY PROFESSION?

Although controversy exists, big data are likely here to stay. It is imperative that we as educators take precautions to ensure the safety of our faculty and our students. There are multiple benefits to big data and we can use the data to inform curricular, instructional, and professional development decisions that are made in our classrooms and in our schools on a daily basis.

When presenting data to any audience, it is extremely important that you use the appropriate figures, tables, and charts to represent the data. There is nothing more embarrassing than sharing data using the wrong type of illustration to tell the story. Depending on the type of data and the story you are trying to tell, you must use the appropriate form of visual illustration. For many people, data can be very difficult to interpret. If you present longitudinal data using a pie chart, you will confuse your audience. This is especially important for educational administrators. You work with schoolwide data on a regular basis. When you want stakeholders to understand your school's story, please make sure you incorporate the appropriate illustrations so your audience can interpret what it is that you are sharing.

CHAPTER 12

Assessing ALL Students

How should performance of special-needs students be evaluated?

AS YOU READ

As you read the chapter, think about the following:

- How do **you** define or explain where ESE kids fit into assessment?
- What is **your** experience with assessing ESE kids?

WHAT ARE THE GENERAL PRINCIPLES OF ASSESSING ALL STUDENTS?

Often, as educators, we create assessments, use end-of-chapter exams, and administer statewide standardized tests to our students to measure their knowledge, skills, abilities, attitudes, and the like. We do not, however, always think about each student's learning style, preferences, and individual needs. I find it interesting that we often think about our students' individual needs when we plan our instruction. We know which students need to sit closer to the front because of vision or hearing limitations. We know which students need us to repeat the information multiple times because of their lack of attentiveness. We know which students require us to closely monitor their behavior because of distractions or movement needs. We know which students require a copy of the teacher's notes to help with interpretations. We know this and we think about this as we plan our lessons. However, we often forget about these individual needs when we plan our assessments.

Typically, we create or use an existing instrument to measure students' knowledge, skills, abilities, attitudes, and so on. We administer that test and then compile the results. Do you ever consider that the students who are easily distracted may need a break during the test? Do you ever realize that students who lack attention may be able to perform better if they are assessed verbally instead of in written form? There are multiple ways to assess student learning—all of which may be different and/or individualized to meet students' needs.

In a previous chapter we learned about error and how various factors can impact test scores. Well, the way you present the test can impact the results. According to Hong and Ehrensberger (2007), presenting information in a variety of ways as well as providing students with real-world examples can help them demonstrate their knowledge and/or skills. They also suggest enlarging fonts and including the test questions on separate pages can help reduce the anxiety for students who get overwhelmed by tests. There are multiple strategies that you may employ in the assessment process. Remember, you use multiple strategies when you present material to differentiate instruction. You should also use differentiation when you assess student learning.

Practices for Varying Assessments to Meet Students' Needs

The following list encompasses strategies for meeting the needs of each student:

- Read instructions aloud.
- Repeat instructions.
- Have students engage in choral reading of instructions.
- Use oral versus written responses to items.
- Reduce the number of items on each page of a test (print on separate pages).
- Increase font sizes.
- Provide real-world examples.
- Use hands-on activities.
- Provide visual models.
- Use multiple test formats.
- Provide verbal prompts for each type of question on the test.
- Allow privacy—reduce distractions by allowing small groups or individuals to test in different location.
- Recommend mental breaks.

Of course, it is imperative that all versions of the test elicit evidence of validity and reliability. Refer to Chapter 6 for collecting evidence of validity and reliability.

WHAT DOES THE LITERATURE SAY ABOUT ASSESSING ALL STUDENTS?

Reading 12.1

MAKING ASSESSMENT AUTHENTIC FOR MULTICULTURAL LEARNERS WITH SPECIAL NEEDS

Tes Mehring

America's classrooms are becoming increasingly diverse. We are becoming the most ethnically, culturally, and linguistically diverse society that has ever existed. Consider the following statistics:

- Today, approximately 40 percent of students enrolled in public schools in the United States are minority students (Gargiulo, 2009).
- Twenty percent of students live in poverty (Gargiulo, 2009).
- Approximately 13 percent of the 48 million students in U.S. schools have special needs that interfere with their ability to learn (O'Donnell, Reeves, & Smith, 2009).
- By the year 2020, multicultural households will account for more than 40 percent of the U.S. population and that percentage will likely grow to 50 percent by 2050 (Gargiulo, 2009).
- By 2020, students of color are projected to make up almost one half of all school-age youth (Gollnick & Chinn, 2009).
- By the year 2050, the U.S. population is projected to be 53 percent Anglo, 24 percent Latino, 13 percent black, 9 percent Asian, and 1 percent Native American (Chinn, 2002).
- About one in five individuals (approximately 19 percent of the U.S. population over the age of 5) speaks a language other than English at home (U.S. Census Bureau, 2007).
- The number of students who are culturally and linguistically diverse and disabled is estimated at approximately 530,000 (Baca, Baca, & Valenzuela, 2004).

The appropriate assessment of all students has been a longstanding concern among special educators for more than three decades. Appropriate assessment is an especially critical issue for children from minority populations. According to Gargiulo (2009), assessment is the "primary vehicle through which access to services is determined and progress is evaluated, using a variety of formal and informal means. Assessment should be a dynamic, multifaceted, multipurpose decision-making process whose primary goal is to evaluate the academic and behavioral progress of a student" (p. 101). The growing number of culturally and linguistically diverse (CLD) students in the nation's classrooms and the tightening of school, district, and state accountability expectations are resulting in a reexamination of the assessment measures we use to refer, diagnose the need for special education services, and document student progress in learning goals. Assessment and instructional practices in American schools were neither created nor designed to be responsive to the range of diversity represented in today's school population. There are several roadblocks to the goal of achieving meaningful and valid assessments of students who are culturally and linguistically diverse. Standardized testing has frequently been criticized for its failure to consider the cultural and experiential background of culturally and linguistically diverse students. Most standardized academic achievement tests require students to have specific culturally-based information in order to perform well on the test (Grassi & Barker, 2010). Many tests are culturally biased toward the mainstream—White, English only, middle to upper class students (Baca et al., 2004).

Many factors must be considered in achieving inclusive and equitable assessment for diverse students. These include the "prior knowledge and language skills that assessment tasks require; whether test content, procedures, or scoring criteria are biased; whether tests are valid for the population being assessed; and whether all students have had the opportunity to learn the material assessed" (Lachat, 2004, p. 67). Assessment must be inclusive, fair, relevant, comprehensive, valid, and yield meaningful information (Gottlieb, 2006). Educators must understand the power of assessment data in helping to provide the evidence that our students are learning and making progress. Ultimately, it is our responsibility to create a bridge, through sound assessment, to ensure the academic success of our students, especially those who come from diverse backgrounds. To large measure, concern about test bias has resulted in efforts to reduce bias through focusing on the test instruments themselves. Many tests were revised in an effort to reduce the number of culturally specific test items (content bias) and the reliance on culturally specific language. Tests were also renormed, or restandardized to reflect the growing diversity of children in schools (Gargiulo, 2009). Even the testing environment and the race of the examiner and his or her interactions with the student have been analyzed. Many school districts, searching for better ways of assessing diverse students are moving more and more toward more authentic, performance-based assessment strategies (Friend & Bursuck, 2006).

CHARACTERISTICS OF AUTHENTIC ASSESSMENTS

Authentic assessment is widely viewed as offering a means to measure student progress on learning standards and as being more responsive to diversity than traditional assessments. When standards are linked to authentic assessment, students must demonstrate their skills and abilities through a range of "performances," and emphasis is placed on student work that involves higher-order thinking and complex problem solving. Students are asked to apply the skills and competence they have learned to "authentic" tasks that represent practical or "real-life" contexts. According to Lachat (2004), authentic assessments have the key features listed below. These assessments

- measure student achievement against a continuum of agreed-upon standards of proficiency;
- emphasize the importance of context through real-life tasks that are "authentic" to the learner;
- focus on higher-order thinking processes and how students integrate information and skills in performing tasks;
- require students to display what they know and are able to do by solving problems (performance tasks) of varying complexity, some of which involve multiple steps, several types of performance, and significant student time; and
- often involve group as well as individual performance on a task.

Ronis (2007) noted that an assessment is considered to be authentic "when it involves students in tasks that are worthwhile, significant, and meaningful; occurs over time (not just one day); is open-ended; and allows for students to demonstrate competence in a variety of ways" (p. 2). Such assessments involve higher-order thinking skills along with the use of a broad range of knowledge. Authentic assessments are brain compatible. They emphasize learning and thinking, especially higher order skills involved in problem solving. Authentic assessments use meaningful tasks that reflect real-life, interdisciplinary challenges. They present students with complex, ambiguous, open-ended problems and tasks that integrate their knowledge and skills. Such assessments usually culminate in student products or performances that recognize and value each student's multiple abilities, varied learning styles, and individual background.

Authentic assessments are more student oriented than traditional assessments and less concerned with the formal characteristics of assessment formats. They also tend to relate more closely both to instruction and to real-life applications of the knowledge and skills involved. Examples of authentic assessments might include the following:

- *Observation* of student behavior in a natural setting (e.g., playground, bus, classroom, cafeteria, etc.) using rating scales, observational recording systems, anecdotal records, or audio recordings
- *Demonstrations/Presentations* that provide opportunities for students to show their learning in oral and media performances or exhibitions
- *Interviews* obtained orally or in writing from significant individuals in a student's life (e.g., parents, teachers, siblings, peers, or the pupil him or herself)
- *Projects and Demonstrations* that provide opportunities for students to show their learning in oral and media performances or exhibitions (e.g., poster, oral or written report, production, skit, speech, Web page, etc.)
- *Work Samples/Permanent Products* that provide evidence of a pupil's classroom performance, typically focused on particular skills (e.g., quizzes, tests, homework, assignments, etc.)
- *Simulations* that provide role-playing tasks that allow students to demonstrate connections among concepts and ability to apply learning in contexts that replicate real-life problem solving
- *Learning Logs/Reflective Journals* in which students reflect on what they have learned and/or their own performance using written or audio recordings
- Portfolios that document a wide range of examples of a student's emerging growth, abilities, and accomplishments over time (e.g., a unit of study, semester, academic year, or multiple years of schooling) through a collection of student work that might include any of the types of authentic assessments presented above. Bohlin, Durwin, and Reese-Weber (2009) indicated that a portfolio is "a systematic collection of students; work that demonstrates accomplishments, growth, and reflection on their learning" (p. 470).

Clearly, one aspect most authentic assessments have in common is that they usually result in a student product or performance. This type of assessment has become increasingly popular because it relates more closely to activities one actually does in life (O'Donnell, Reeve, & Smith, 2009).

ADVANTAGES OF AUTHENTIC ASSESSMENT

O'Donnell, Reeve, and Smith (2009) cited several advantages supporting the use of authentic assessments. These assessments are directly related to the skill being assessed. For example, the best way to determine whether students can write legibly using cursive or manuscript is to collect a permanent product that provides samples of student handwriting. Authentic assessments cause students to prepare for the assessment in ways that are conducive to learning. Authentic assessments also tend to be more engaging for students than standardized assessments. Lachat (2004) cites several additional advantages for authentic assessments with diverse learners. These assessments promote active student learning and support instruction. They assist teachers to make instructional decisions by actively involving both students and teachers in the learning process. Authentic assessments minimize the likelihood of drawing conclusions from limited performance opportunities. They offer children from different backgrounds varied ways to display their knowledge and abilities. These assessments provide information that can be used to form a profile of a student's individual strengths and weaknesses. Authentic assessments allow teachers to document the broad-based process of learning, follow children's development, and create differentiated profiles of student's accomplishments. Teachers and students often collaborate in the learning process. These assessments allow students to display proficiencies in a wide variety of ways, allow more dynamic approaches to assessing student learning, and allow developmental learning to be profiled over time. Meisels, Dorfman, and Steele (1995) suggest that authentic assessments provide teachers with evidence to monitor student progress over time and to adjust instruction appropriately. Farr and Trumbull (1997) stated the following advantages:

> Because authentic assessments involve multiple ways of demonstrating proficiency, they invite students to draw on multiple intelligences and to display varied cognitive and communicative styles. As a result, these assessments provide a wider range of opportunities for students to show what they know and can do. Because they make greater cognitive demands on students than traditional standardized tests, authentic tasks invite a fuller range of responses, provide a richer picture of what students have learned, and allow for the ongoing assessment of their higher-order thinking skills. (p. 89)

Lachat (2004) indicated that the flexibility of authentic assessments allows teachers to vary the methods used to diagnose the learning of students whose cognitive and cultural styles may cause them to perform poorly on standardized tests. "By offering a range of contexts—including opportunities to work alone, in pairs, or in groups—teachers can vary assessment settings to reflect cultural preferences and also to evaluate the impact of these contexts on particular students' progress. By examining how students solve tasks, teachers can better differentiate between learning problems caused by poverty or language competence, and those caused by limited content knowledge" (p. 89). As it appears, authentic assessments are brain compatible. Ronis (2007) stated that

> When correctly designed, they emphasize learning and thinking, especially those higher-order thinking skills involved in problem solving. Authentic assessments comprise meaningful tasks that reflect real-life, interdisciplinary challenges; they present students with complex, ambiguous, open-ended problems and tasks that integrate their knowledge and skills. Such assessments usually culminate in student products that integrate their knowledge and skills. Such assessments usually culminate in student products or performances that recognize and value each students' multiple abilities, varied learning styles, and individual background. (p. 13)

DISADVANTAGES TO AUTHENTIC ASSESSMENTS

There can also be disadvantages to authentic assessments. These tools can be difficult to objectively score. Bias can also enter into scoring of these assessments even if criteria are explicit. For example, a general classroom teacher who is frequently challenged by disruptive behavior from an African American student with behavior disorder, may inadvertently or overtly score his orally presented speech about early settlers in Alaska lower than peers who tend to be compliant and engaged in class sessions even though a scoring rubric is very explicit. Performance anxiety might also be a problem in authentic assessment because some students are very reluctant to engage in any kind of performance in front of their peers. If authentic assessments involve work conducted outside the classroom, it can be difficult to know if the student completed the assessment activity independently, or if they had assistance from family members or even friends. Finally, authentic assessment can be very time consuming.

CONCLUSION

Despite criticisms that standardized tests do not always assess what students are learning, and that their emphasis is on mostly factual knowledge rather than higher-order thinking and application, Burke (2009) indicated that they are still the "yardstick that the public and policymakers use to measure educational progress. Standardized tests are viewed by many people as being valid and reliable and, for the most part, the most effective method to compare students, schools, districts, states, and countries" (p. 1). Hargreaves and Shirley (2008) indicated that the data on these existing assessment foci and the economic need for increased innovation and creativity has necessitated a shift in education reform. Although research-based instructional strategies and classroom management strategies are critical components of teaching, the research on authentic assessment indicates that it is one of the major factors that improve student achievement. As

Marzano (2006) pointed out, "To the surprise of some educators, major reviews of the research on the effects of classroom assessment indicate that it might be one of the most powerful weapons in a teacher's arsenal" (p. 2).

Authentic assessments provide feedback for teachers and students, evaluate students' knowledge and understanding of key concepts and standards, and guide the instructional process by differentiating teaching to meet the diverse needs of all learners. Authentic assessments that are frequent and integrated seamlessly with instruction provide a continuous feedback loop that informs instruction. Stiggins (2004) concluded that the educational system will continue to use both standardized tests and authentic assessments and both will continue to provide valuable information for important decision making. Clearly, authentic assessments are valuable teaching tools that should be used to promote meaningful learning for all students. Most advocates of authentic assessment do not want to totally eliminate traditional tests. What they seek is a good mix of assessment practices. As educators, we are constantly challenged to make informed decisions about our students. We plan, gather, and analyze information from multiple sources over time so that the results are meaningful to teaching and learning. If reliable, valid, and fair for our students, assessment can be the bridge to educational equity. Using multiple assessment measures in multiple contexts to obtain a more accurate assessment of student's proficiency levels provides a more accurate interpretation of functioning. Authentic assessments provide an option for assisting educators to accomplish this goal.

REFERENCES

Baca, L., Baca, E., & Valenzuela, J. (2004). Background and rationale for bilingual special education. In L. Baca & H. T. Cervantes (Eds.), *The bilingual special education interface* (4th ed., pp. 1–23). Upper Saddle River, NJ: Pearson Education.

Bohlin, L., Durwin, C. C., & Reese-Weber, M. (2009*). EdPsych modules*. Boston: McGraw Hill Higher Education.

Burke, K. (2009). *How to assess authentic learning* (5th ed.). Thousand Oaks, CA: Corwin Press.

Chinn, P. (2002). *Changing demographics in America.* Nashville, TN: Alliance Project, Vanderbilt University.

Farr, B. P., & Trumbull, E. (1997). *Alternate assessments for diverse classrooms*. Norwood, MA: Christopher-Gordon.

Friend, M., & Bursuck, W. (2006). *Including students with special needs* (4th ed.). Boston: Allyn and Bacon.

Gargiulo, R. M. (2009*) Special education in contemporary society* (5th ed.). Los Angeles, CA: SAGE.

Gollnick, D., & Chinn, P. (2009). *Multicultural education in a pluralistic society* (8th ed.). Upper Saddle River, NJ: Pearson Education.

Gottlieb, M. (2006). *Assessing English language learners: Bridges from language proficiency to academic achievement*. Thousand Oaks, CA: Corwin Press.

Grassi, E. A., & Barker, H. B. (2010). *Culturally and linguistically diverse exceptional students: Strategies for teaching and assessment*. Thousand Oaks, CA: Sage.

Hargreaves, A., & Shirley, D. (2008). Beyond standardization: Powerful principles for improvement. *Phi Delta Kappan, 90*(2), 135–143.

Lachat, M. A. (2004). *Standards-based instruction and assessment for English language learners*. Thousand Oaks, CA: Corwin Press.

Marzano, R. J. (2006). *What works in schools: Translating research into action*. Alexandria, VA: Association for Supervision and Curriculum Development.

Meisels, S. J., Dorfman, A., & Steele, D. (1995). Equity and excellence in group-administered and performance-based assessments. In M. Nettles & A. Nettles (Eds.), *Equity and excellence in educational testing and assessment*. Boston, MA: Kluwer Academic.

O'Donnell, A. M., Reeve, J., & Smith, J. K. (2009). *Educational psychology: Reflection for action*. Danvers, MA: John Wiley & Sons, Inc.

Ronis, D. (2007). *Brain-compatible assessments* (2nd ed.). Thousand Oaks, CA: Corwin Press.

Stiggins, R. J. (2004). *Student-involved assessment for learning*. Upper Saddle River, NJ: Merrill/Prentice Hall.

U.S. Census Bureau. (2007). Statistical abstract of the United States: 2008 (127th ed.). Washington, DC: Author.

HOW DOES ASSESSING ALL STUDENTS APPLY TO MY PROFESSION?

Maximizing student learning and success should be our goals. So, what can we do to maximize student learning and success? Make sure that when we assess our students, we use assessment strategies that meet their learning needs. The presentation of the test itself can impact the scores. The environment and/or testing conditions can impact results. So, as educators, it is our responsibility to ensure the comfort level of our students, suitable seating arrangements, appropriate temperatures, effective lighting in the room, and most importantly, the format of the test aligns to the individual student's needs. Yes, this requires more time and effort for the teacher. But, if done effectively, students' performance will better reflect accurate results.

CHAPTER 13

Teacher, Leader, and Policy: Considerations and Implications

What considerations and implications in policymaking?

AS YOU READ

As you read the chapter, think about the following:

- How do **you** define or explain various considerations?
- What is **your** experience with various considerations?

WHAT ARE THE GENERAL PRINCIPLES OF POLICY CONSIDERATIONS AND IMPLICATIONS?

What Are the General Principles?

Both classroom teachers and educational leaders are faced daily with policies that affect their practices, decision making, funding, curriculum, instructional strategies, and assessment practices. Often, educators argue that they are the best resources for making decisions about policies because they know firsthand about the implications of those policies. However, politicians and policymakers may make assumptions about what is best for education and its stakeholders. Educational planning and policymaking processes have always been linked. However, Haddad and Demsky (1995) reported a noticeable shift in educational planning during the 1980s and 1990s. Members of this group concluded that the planning phase is now intertwined with the policymakers. The members questioned the potential constraints and implications of the policymaking process will have on the educational planners.

Let us take a look at one example of a policy that has had significant implications, perhaps some good and some bad. Think about the implications of value-added modeling, a policy with implications of high stakes. Econometric approaches to measuring the value a teacher adds may seem straightforward. However, researchers argue that measuring added value cannot be easily quantified because the hierarchical nature of educational data violates the assumption of independence inherent in single-level data. Furthermore, value-added models cannot possibly take into account the complexity of measurement error (Marsh et al., 2011). Reflecting on the measurement concepts explored in this book may give you pause as you think about using student achievement, the dependent variable, as the single measure of effectiveness. Past research has demonstrated that the actual test used will impact students' scores (Bill &Melinda Gates Foundation, 2010; Corcoran, 2010). As we previously discussed, test results are a mere construct for which we make conclusions. These scores are not exact measures of knowledge. Thus, although the intent of determining effectiveness based on student achievement results may be a worthwhile goal, it cannot be the means to an end. An opposing argument, however, is that the primary purpose of education is to increase students' knowledge and skills. Thus, student achievement is the best way to determine what students know and can do.

I used value-added modeling to measure effectiveness as an example in this brief review of implications of policies on educators. Decision making is a critical part of policy processes. And clearly, decision making related to a policy has been preceded by various analytical and/or political activities. Furthermore, those decisions are followed by planning activities for implementation of the policies. As indicated by Haddad and Demsky in the Education Policy-Planning Process: An Applied Framework (1995), "A policy decision in any one component can have strong repercussions throughout the system. Externally, education seems to be everyone's business and nearly everyone feels qualified to have an opinion about it. Policy making, therefore, involves balancing a number of contradictory demands, and soliciting support, or at least tolerance, from the many different segments of society that have an interest in education" (p. 23).

WHAT DOES THE LITERATURE SAY ABOUT POLICY CONSIDERATIONS AND IMPLICATIONS?

Reading 13.1

SELECTIONS FROM *EDUCATION POLICY-PLANNING PROCESS: AN APPLIED FRAMEWORK*

Wadi D. Haddad and Terri Demsky

This booklet analyzes educational policy making as a cornerstone of (Fundamental) educational planning. The policy analysis framework and case studies presented here provide planners with both a conceptual and an operational guide for understanding the critical linkages in the policy-planning process for education.

Chapter I offers an integrated model of educational decision making that emphasizes the role of the formal policy-making process (and its analytical rationality) within the context of the key policy actors (from the administrative and political context). To capture the details of the decision-making process itself, an analytical framework is presented that goes beyond the initial decision point to examine both the preceding actions (contextual assessment, technical analysis, and the generation, valuation, and selection of policy options) and the subsequent activities (planning and conducting implementation, impact assessment, and, where appropriate, remediation and redesign). Thus, the framework covers the full policy-planning process but with a focus on the facilitating and constraining effects that policy decisions (and how they were derived) can have on the choices available to educational planners.

In Chapter 11, the framework is applied to four exemplary case studies of the educational policy cycle. The four cases (Peru, Jordan, Thailand, and Burkina Faso) were selected because sufficient data (and elapsed time) exist on each to allow discussion of the full policy cycle; also, the geographical, economic, pedagogical, and political diversity of the examples emphasizes that the policy framework presented here is generic and not limited in applicability by any of these factors. Chapter 111 summarizes the lessons from and across the case studies, and identifies the elements in the policy-planning process that appear to have contributed to the success or failure of policy reforms. Chapter IV summarizes the implications for planners.

This analysis should be of value to educational planners in two major ways. First, the methodology of the framework and conclusions of the case studies should help in the analysis of current educational policies and decision-making procedures (an analysis of policy). Second, the framework can be applied to the evaluation of proposed policies and used to forecast policy outcomes and the probability of successful implementation, given the country context of fiscal and managerial capacity, political commitment, etc. (an analysis for policy). Planning that is not based on a solid understanding of educational policy making will fail; it will fail not primarily because of any technical planning errors but because the planners did not understand why and how these policies evolved and how planning results should lead to new cycles of policy analysis and formulation.

FRAMEWORK FOR EDUCATION POLICY ANALYSIS

The notion of educational planning—making the education sector grow and function more effectively—may implicitly suggest a well structured field of unambiguous issues, clearly defined objectives, mutually exclusive choices, undisputed causal relationships, predictable rationalities, and rational decision-makers. Accordingly, sector analysis has predominantly focused on the content—the 'what' of educational development: issues, policies, strategies, measures, outcomes, etc. In contrast to this simplistic vision, educational planning is actually a series of untidy and overlapping episodes in which a variety of people and organizations with diversified perspectives are actively involved—technically and politically. It entails the processes through which issues are analyzed and policies are generated, implemented, assessed and redesigned. Accordingly, an analysis of the education sector implies an understanding of the education policy process itself—the 'how' and 'when' of educational development. The purpose of this section is to suggest a scheme or series of steps through which sound and workable policies can be formulated, and then, through effective planning, put into effect, evaluated and redesigned.

Policy Definition and Scope

Since the policy process is a crucial element in educational planning, it is essential to clarify the concepts of 'policy' and 'policy making' before proceeding any further. Understandably, competing definitions of 'policy' are numerous and varied. For the purposes of this paper, policy is defined functionally to mean: *An explicit or implicit single decision or group of decisions which may set out directives for guiding future decisions, initiate or retard action, or guide implementation of previous decisions*. Policy making is the first step in any planning cycle and planners must appreciate the dynamics of policy formulation before they can design implementation and evaluation procedures effectively.

Policies, however, differ in terms of their scope, complexity, decision environment, range of choices, and decision criteria. This range is schematically depicted in *Figure 13.1*. Issue-specific policies are short-term decisions involving day-to-day management or, as the term implies, a particular issue. A programme policy is concerned with the design of a programme in a particular area, while a multi-programme policy decision deals with competing programme areas. Finally, strategic decisions deal with large-scale policies and broad resource allocations. For example:

Strategic: How can we provide basic education at a reasonable cost to meet equity and efficiency objectives?

Multi-programme: Should resources be allocated to primary education or to rural training centres?

Programme: How should training centres be designed and provided across the country?

Issue-specific: Should graduates of rural centres be allowed to go into intermediate schools?

Another example:

Strategic: Should we or do we need to introduce diversified education?

Multi-programme: How should we allocate resources between general education, vocational education, and diversified education?

Programme: How and where should we provide diversified education?

Issue-specific: How should practical subjects be taught in diversified schools?

Obviously, the broader the scope of a policy is, the more problematic it becomes. Methodological and political issues become more pronounced such as, definition of the problem in conflictive societies; use of analytical techniques and optimization; questions of proper theoretical base, measurement, valuation and aggregation; hard objective data vs. soft subjective data; and technical analysis vs. public participation. For a more detailed treatment of this subject, refer to Michael Carley (1980).

Issue-specific *Programme* *Multi-programme* *Strategic*

	Complexity	
Low	<----------------------------------->	High
	Decision Environment	
Precise	<----------------------------------->	Imprecise
	Number of Alternatives	
Low	<----------------------------------->	High
	Decision Criteria	
Narrow	<----------------------------------->	Broad

FIGURE 13.1 **Policy scope**

Policy Making

The term 'policy making' like 'policy' implies competing conceptions and assumptions. A study of the theoretical and empirical work of social scientists reveals the two essential dimensions of policy making: who does it (the actors) and how (the process). Historically, the actor in policy making has been considered unitary and rational; more recently policy analysts have introduced the organizational (public interest) model and the personalistic (self-interest) model. The process element has fluctuated between a synoptic (comprehensive) approach and an incremental approach.

C. Lindblom and D.K. Cohen (1979) laid out the differences between the synoptic and incremental methods of policy making. According to him, the synoptic method entails, in its extreme form, one single central planning authority for the whole of society, combining economic, political, and social control into one integrated planning process that makes interaction unnecessary. It assumes: (a) that the problem at hand does not go beyond man's cognitive capacities and (b) there exist agreed criteria (rather than social conflict on values) by which solutions can be judged and (c) that the problem-solvers have adequate incentives to stay with synoptic analysis until it is completed (rather than 'regress' to using incremental planning).

Incremental policy making, on the other hand, relies on interaction rather than on a complete analysis of the situation to develop a blueprint for solving problems. The incremental approach to policy making is built on the following assumptions: (a) Policy options are based on highly uncertain and fluid knowledge, and are in response to a dynamic situation (everchanging problems, and evolving contexts); (b) No 'correct' solution can therefore be found, or technically derived from a diagnosis of the situation. Thus, no sweeping or drastic reforms should be attempted; (c) Only incremental and limited policy adjustments can be made; and (d) Policy adjustments are expected to remedy an experienced dissatisfaction with past policies, improving the existing situation or relieving an urgent problem. Consequently, these adjustments should be tentative—and in some cases temporary—and must be revised as the dynamics of the situation evolve.

G. T. Allison (1971) developed two alternative models to the commonly assumed model of the unitary rational policy-maker: (a) the Organizational process model, and (b) the governmental politics model. The first model assumes a complex government consisting of a conglomerate of semi-feudal, loosely allied organizations, each with a substantial life of its own. Decisions are based on the output of the several entities, functioning independently according to standard patterns of behaviour but partially co-ordinated by government leaders. The second model carries this concept further. While it also assumes an organizational approach to decision making, the Governmental Politics model plays up the part of individuals in the process. Government decisions are not made by a monolithic state based on rational choice, but rather are negotiated by various leaders who sit on top of the organizations involved in that particular decision-making process. Each leader is compelled by his own conception of the problem as well as by the imperatives of his organization and his own personal goals.

A consolidated model for policy making. Neither of the two dimensions alone (process and actors) fully captures the dynamics of policy making. They need to be combined and restructured into a different configuration, as the topography in Figure 13.2 illustrates.

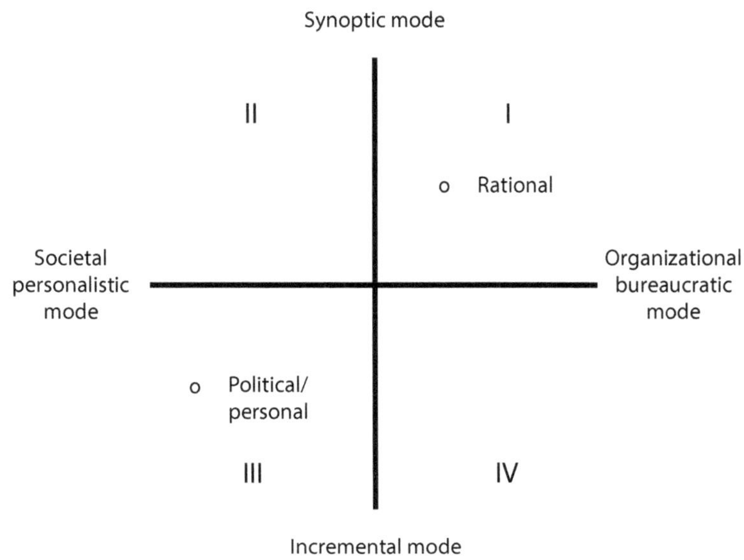

FIGURE 13.2 **Dimensions of policy making**

The actor in policy making is placed on the horizontal-axis—at one end of the spectrum is the societal/personalistic mode, wherein decisions are reached by negotiation among a variety of interest groups (including government ministries, teachers' unions, etc.), driven by their own conception of the problem and individual values. On the other end is the organizational/bureaucratic mode wherein decisions are made within the organizational entity (i.e. the military, the international community, etc.). The process of policy making—from the incremental to the synoptic approach—is placed on the vertical-axis. These two dimensions generate a new topography.

On the one extreme of this new topography (in quadrant I) is the rational model which is a composite of the synoptic method and the organizational/bureaucratic mode. Decision making at this extreme is unitary, rational, centrally controlled, completely technical and value maximizing. On the other extreme (in quadrant III) is a composite of the incremental method and the societal/personalistic mode. Policy making here is a political activity characterized by self-interest, political bargaining, value judgement and multiple rationalities. One can easily argue that most policy making falls somewhere between these two extremes. Analytic techniques caried on in ignorance of political, social and bureaucratic realities do not go very far. Similarly, a pattern of vague and unsystematic political decisions loaded with self-interest, patronage and value judgements can lead to breakdown, if not to chaos.

In sum, a balanced perspective of policy making places analytical rationality within the context of political and institutional aspects of policy making. This is in line with Douglas North's pioneering work on institutional economics that gained him the Nobel Prize in 1993. In addition to modifying the rationality postulate, he extended the economic theory by incorporating ideas and ideologies into the analysis and allotted a fundamental role to institutions for societal change: they are "the underlying determinant of the long-run performance of economies" (North 1990).

This balanced view of policy making is most appropriate for education. Studies of educational policy making all point to the complexity and multifaceted character of this process due to the nature of both the educational system and the educational change. One of the more important characteristics of the education system lies in its salient linkages with the socio-economic structure. Any policy changes, therefore, are not purely technical but have socio-political-economic dimensions. For instance, any attempt to modify the system, which is perceived by one group or another as lowering the chances of their children to progress socially or economically, will meet with strong opposition. Therefore, the whole notion of reform for democratization is essentially a political issue. Another complex set of linkages exists between the education system and the economy, whereby the school is seen as the solution to a wide range of economic problems. This belief is the source of much of the impetus for policy changes.

Internally, the educational system is an intricate network of institutions interlocking horizontally and vertically. A policy decision in any one component can have strong repercussions throughout the system. Externally, education seems to be everyone's business and nearly everyone feels qualified to have an opinion about it. Policy making, therefore, involves balancing a number of contradictory demands, and soliciting support, or at least tolerance, from the many different segments of society which have an interest in education.

Conceptual Framework for Policy Analysis

Although decision making is a crucial event in the policy process, clearly it is preceded by analytical and/or political activities (analysis, generation of options, bargaining, etc.) and followed by equally important planning activities (implementation, assessment, and possible redesign). This booklet introduces a framework for education policy analysis that covers the pre-policy decision activities, the decision process itself, and the post-decision planning activities. This framework is not a description of actual activities, but rather a conceptual model to extract and specify those elements that can be detected and analyzed. It therefore should be broad enough to capture and integrate the intricate process of any policy making model (Figure 13.2), yet at the same time it should disaggregate the process into components to determine how they work and interact. The resultant framework, summarized schematically in Figure 13.3 and discussed in detail below, consists of seven policy-planning processes, the first four of which deal with policy making, the fifth with planning and sixth and seventh with policy adjustment:

(I) Analysis of the existing situation.
(II) The generation of policy options.
(III) Evaluation of policy options.
(IV) Making the policy decision.
(V) Planning of policy implementation.
(VI) Policy impact assessment.
(VII) Subsequent policy cycles.

This framework looks complicated because, inevitably, it is multifaceted and covers a wide range of processes. However, any attempt to restrict policy analysis to certain elements or to disregard one element results in an incomplete approach to policy analysis, and leads to the historical controversy of the rational vs. the political, or the bureaucratic vs. the organizational approaches in the literature and in public debate. The above seven elements of the framework will be used to frame the case studies in Chapter II and the conclusions in Chapter III.

A. Analysis of the Existing Situation

A policy change is normally a response to a problem or set of problems in the sector, and must, therefore, start with an appreciation of the educational sector and its context. In addition to the analysis of the sector itself, policy analysis should consider a number of aspects of the social context, including political, economic, demographic, cultural, and social issues which are likely to affect the decision making and even implementation processes of the education sector.

Country Background

The general character of a country (location, geography, population, culture, and social stratification patterns) has obvious implications for education policy analysis. This makes the process of educational policy making more difficult and in a number of ways. Typically different groups have different values about the role of education. Insofar as education represents access to economic and political power, then different access or interest in education also means differential access to power. Resulting conflicts and struggles are particularly acute in countries where the distribution of access to goods and services has become increasingly unequal.

Political Context

The preceding observation emphasizes that an analysis of the political environment is necessary for an understanding of the national decision-making process, the comparative value of education, and the role that education must play in the socio-political process. It is worth distinguishing between the priorities of the national political elite relative to development and those of the educational elite relative to education. It is not only that the head of the ministry of education may have different plans from those of the political elite which appointed him but in many countries there is considerable autonomy provided to the educational sector. It is not at all unusual for the two sets of objectives to be at odds with each other or at least not to be tightly intertwined.

The capacity of the state to do planning at the national level is another critical variable in the institutional analysis of the political sector. Moreover, the professional background of the bureaucrats who do the policy planning and where they have been trained (perhaps in foreign universities) can also affect the ideology of the elite.

Finally, the institutional structure of the political sector has implications for educational development. Many developing countries do not have well developed parties but if they do, their values and preferences need to be calculated as part of the analysis of the political context.

Economic Context

In this area, the analyst wants to understand the present macro-economic situation in general and the human resources situation in particular. It is, however, more important to estimate the likely trends in the various sectors in the future and the financial resources of the country in order to assess what the economy requires from the education sector and what the sector expects to face from the rest of the economy, particularly in terms of general infrastructure and financial resources.

First, variables such as demographic shifts, urbanization, and migration, coupled with the likely growth in various sectors of the economy, will have a significant impact on labour markets and consequently on needs for education and skill training. Second, the level of economic development will set enormous constraints on the capacity of the educational system to build schools

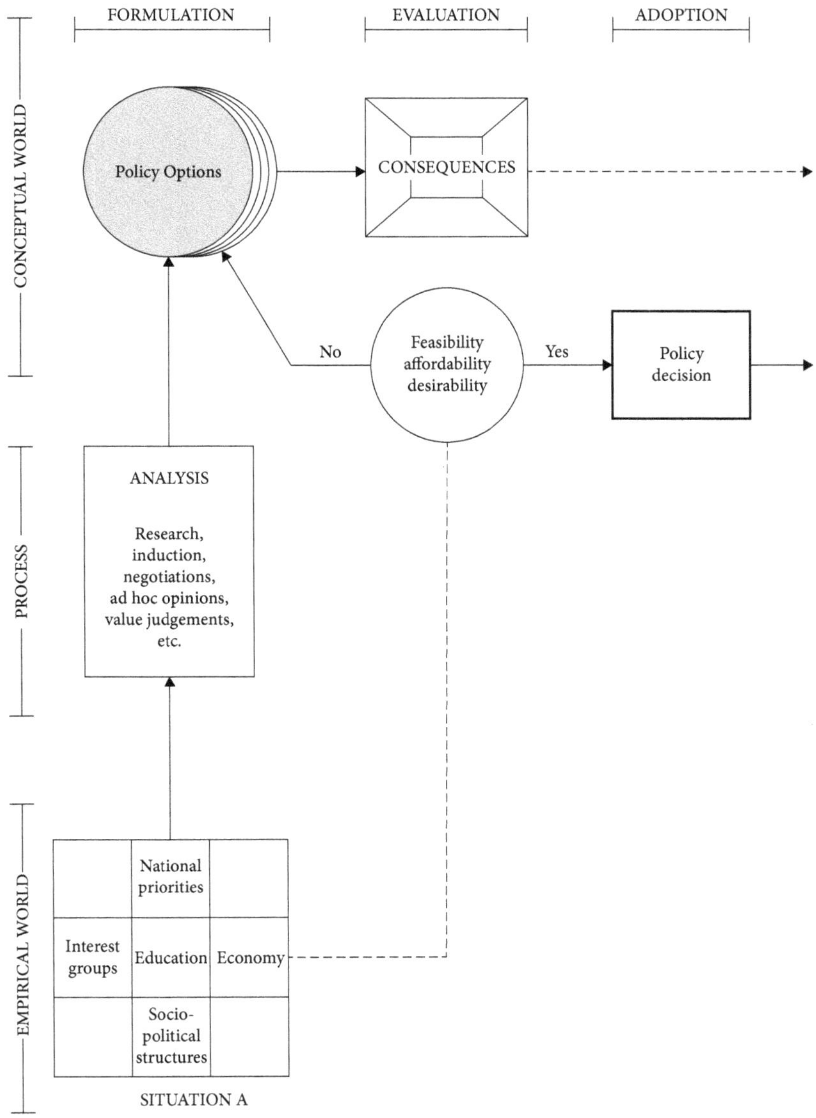

FIGURE 13.3 **Conceptual framework for policy analysis**

and to expand. It is difficult to build schools without the necessary economic infra-structure to say nothing of the presence of firms with the necessary capabilities. The level of economic development also sets the range of possible taxation by the government, which in turn will influence educational expenditures. Third, the economic growth rate is important not only for estimating the likely need for certain kinds of skills but also for estimating the future amounts of slack resources. This is necessary because, as the rate of growth increases, more funds are often made available to education; by the same token, as it decreases, allocations to education are among the first cut.

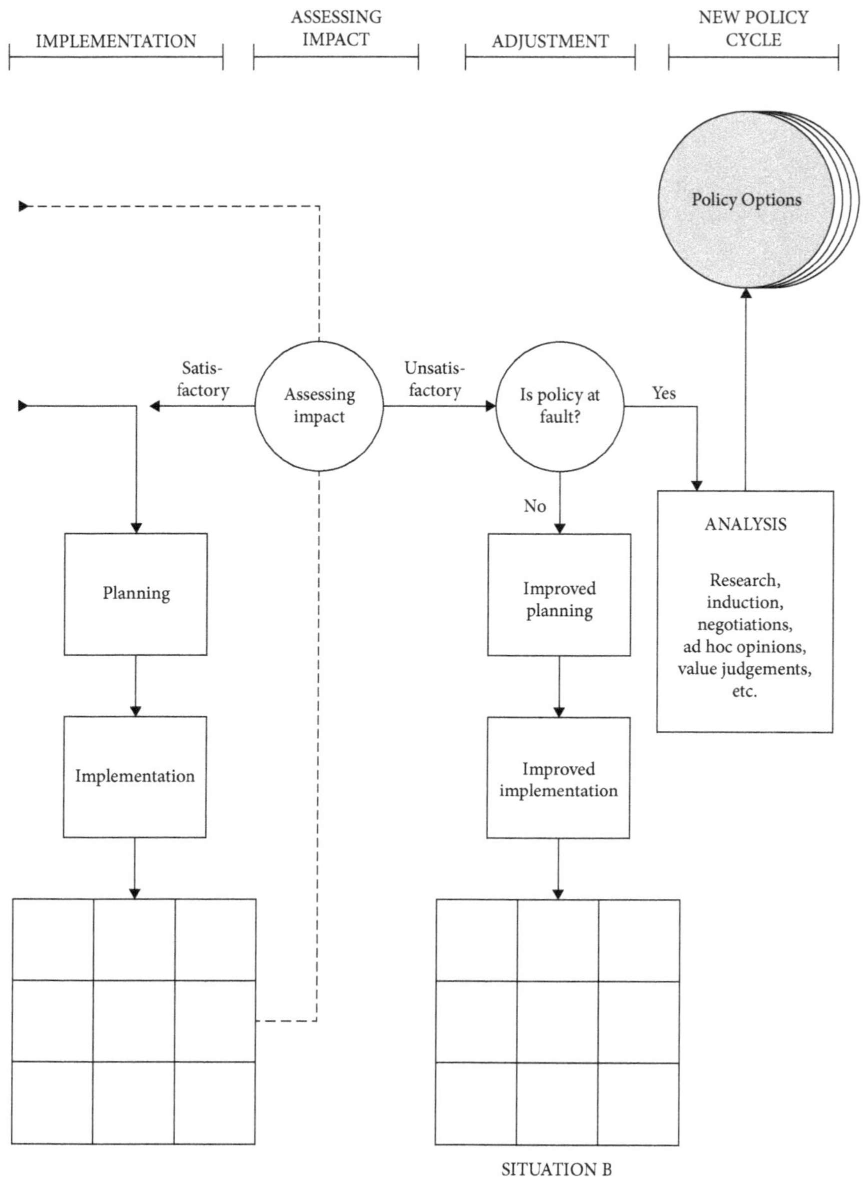

FIGURE 13.3 ***(Continued)***

Education Sector

Sector analysis starts with an identification and understanding of the major sectoral issues relevant to the country. These issues may be explored under six categories: (i) access to educational opportunities; (ii) equity in the distribution of educational services; (iii) structure of the education system; (iv) internal efficiency; (v) external efficiency; and (vi) institutional arrangements for the management of the sector. For a full description of these components and the analytic techniques utilized refer to Haddad and Demsky (1994), Kemmerer (1994), Coombs and Hallak (1987), Mingat and Tan (1988), and Windham (1988a, 1988b).

An analysis of the above issues should take into consideration their evolutionary nature: how have issues in the development of the educational system changed over time? The meeting of one educational need or solution of one problem frequently creates another. For example, the expansion of the system and the provision of new facilities naturally lead to issues about the quality of the education provided and the capacity of the educational administration to handle a larger educational system. In addition, the analysis of education across time can sensitize one to the tendency for the system to oscillate between objectives which are somewhat incompatible. A historical and evolutionary perspective on the dynamics of policies across time allows the analyst a better sense of why a particular policy is being advocated at the moment. By studying the past, one also learns the likely speed with which educational policies can be implemented.

Dynamics of Change

An assessment of the present situation cannot be complete without evaluating the forces for or against change in the event that policy changes need to be made. Such an assessment has implications for the chances of success of different types of policies and for strategies that must be employed to promote and implement such policies. Nobel laureate North (1994) confesses that "there is no greater challenge facing today's social scientists than the development of a dynamic theory of social change". Meanwhile, he asserts that "individuals and organizations with bargaining power as a result of the institutional framework have a crucial stake in perpetuating the system". One key socio-political factor to analyze, therefore, is the presence and relative strength of interest groups.

In developing countries it is impossible to specify what might be all the relevant interest groups but at minimum one can start with the providers of education, most notably, teachers, and the consumers, most notably, parents, students and employers. If the former are well organized—they often are—they can be a powerful force in supporting or opposing any educational change. Their interests are likely to be threatened if the educational change results in some challenge to their status or prerogatives. The consumers can also be powerful, but are generally fragmented. They may be divided into different cultural or occupational, or socio-economic groups. Frequently, these will have quite different interests in both the quantity and quality of education. Consumer groups most closely connected with either political policy-makers or decisionmakers within the education system will be able to exert disproportionate influence. In addition, consumers who can organize themselves into forceful street demonstrations, as have some university students, can effect policy changes very favourable to themselves. Therefore, policy planners need to identify interest groups and assess their openness to reform. For those interest groups identified as anti-reform, planners need to determine how well organized they are, how much power they have in society, and how willing they are to exercise their power.

A separate interest group comprises the officials who adrninister an education system. Studies suggest that bureaucrats find it in their self-interest to maintain a moderate expansion of the educational system. They tend also to value whatever configuration of education is current and to resist policies that would alter it. Therefore, one important element in policy analysis is to understand what the self-interests of the educational bureaucrats are and to recognize that these are not necessarily identical with those of the teachers, other educational professionals or consumers. Finally, the pressure to see change happen can come from individuals or groups outside the education sector (as in the case of Peru, described in Chapter II) or from external actors, both individual experts and development agencies (as is the case of Burkina Faso).

B. The Process of Generating Policy Options

New policies are usually generated when the present situation of the sector and its context is perturbed by a problem, a political decision or a reorganization scheme (overall national planning). Policy options can be generated in several different ways to accommodate the disequilibrium. For analytical purposes one can group these processes under the following four modes: systemic, incremental, ad hoc and importation. In concrete situations, though, several of these modes may be combined.

The Systemic Mode

The title may suggest that this is the preferred or best method for generating policy options. This is not necessarily the case because under certain conditions this mode may prove to be defective or impractical. The systemic mode is characterized by three operations: generation of data, formulation and prioritization of options, and refining options. Data are usually derived from two sources: sector analysis, and the existing body of professional knowledge (conventional wisdom, research synthesis, comparative indicators, etc.).

Formation of options under this mode is a fairly complicated process of induction. If based on data alone, a large number of options can be generated to fit the different 'givens' of the sector and its context. At its extreme, intellectual induction seeks to anticipate all possible policy outcomes by thinking through all possible contingencies. It then proceeds to identify optimal or at least efficient options. However, a variety of intellectual, political, social and professional constraints limit the range of policy options. Moreover, options may be given different weights and priorities depending on the perceived importance of the sectoral issues, the relative strength of the interest groups, and the possible combination of different options.

Some of the policy options may be subjected to a microcycle of problem identification: policy formulation—verification—modification or retention. This is a blend of induction and sequential interaction. The experimentation or pilot studies approach adds an input into the data base and to the 'weighting' of the policy options.

The Incremental Mode

Once a problem within the educational system is recognized, then a solution is frequently forced upon the system. This is especially likely to occur when there is a public debate about a problem. Given widespread interest and discussion, the educational system is forced to do something to maintain its legitimacy. The sense of urgency necessitates a quick response. Since the problem is likely to be located in one particular segment of the system, then the issue is how to formulate a policy to adapt the system to the response. This is sometimes called the 'acting out' approach whereby the policymaker seeks to adjust present difficulties rather than to anticipate future ones, thereby promoting incremental improvements.

The Ad Hoc Mode

Sometimes the problem is outside the educational system. It may not even be a problem but instead the emergence of a new elite or a major political event which requires that the educational system make some adjustments or changes. Here the policy may have no rational basis within the education sector.

The Importation Mode

There are many innovations and fashions in educational systems around the world. These can be the source of the policy options considered. Foreign specialists, operating as consultants for international agencies, can provide the stimulus for this mode.

However, a certain policy adopted elsewhere can be imported successfully only if it meets the needs of particular groups in the society, i.e. if there is an importer.

C. Evaluation of Policy Options

Policy options can be evaluated only if alternative scenarios are developed to allow estimations of the likely implications of the options considered. The 'imaginary' situation that would be created if a policy option were implemented is compared with the present situation, and the scenario of transition from the existing to the imaginary case is evaluated in terms of desirability, affordability, and feasibility.

Desirability

This involves three dimensions: (1) The impact of the option on the various interest groups or stakeholders: who would benefit? who might feel threatened? how might the potential losers be compensated? what would make the option desirable to all stakeholders? (2) compatibility with the dominant ideology and targets of economic growth articulated in national development plans; and (3) in some cases, the impact of a policy option on political development and the stability.

Affordability

The fiscal costs of the change as well as the social and political costs need to be evaluated. The difficulty of making these estimations lies in the ability to predict future trends, including economic growth. This is especially important because educational expenditures are more vulnerable to changes in economic situations and political objectives than some other kinds of public expenditure. Therefore, alternative economic scenarios need to be considered. Further, private costs (will a reform require consumers to share the costs, and if so what happens to the poorer groups?), opportunity costs (are there other measures which might benefit the education system, but would have to be foregone to pay for the current proposal?) and political costs (if an option favours one group over another, is the government willing to pay the political cost?) should also be weighed.

Feasibility

Another and very different kind of implication is the availability of human resources for implementing the change. Fiscal resources are easy to compute. More difficult is the estimate of what level of training is required of teachers (the more sophisticated the programme and/or technology involved, the more highly trained the personnel need to be) and whether there are enough personnel to implement the policy option. In many developing countries, highly trained personnel may be in short supply. This then raises the question of whether they can be imported or trained and at what cost. Equally important is the presence of the institutional culture (norms, procedures, environment) necessary to attract, retain, and effectively utilize trained personnel in transforming policies into plans and implemented programmes. Another element in the calculus of feasibility is time. Most studies of education projects indicate that there are frequent time overruns in implementation. More realistic estimates of time need to be made and can only be done by the careful assessment of the implementation capabilities and experiences.

The issue of sustainability should fare prominently when the above criteria are applied. Education initiatives have to be sustained politically and financially over a lengthy period of time to reach fruition. To ensure that, the long-term implications of policy options should be weighed within an overall sectoral policy, itself embedded in a prudent macro framework, and consistent with long-term national aspirations.

D. Making the Policy Decision

Rarely would a policy decision be the considered consequence of the evaluation and previous stages of the decision process—the culmination of a process during which all information relevant to the decision was gathered and carefully analyzed so that a totally optimal policy might be designed and selected. The variety of conflicting interests and rationalities requires that the policy which is selected engineers 'trade-offs' among these interests. The resulting policy may not be optimal for any single interest group, but such a bargained result is necessary to have the broad base of political support which will be needed to take the policy from the drawing board to implementation. In addition, political pressures, oversights in evaluation, or the simple pressure of time may short-circuit the process. A minister with a 'pet idea', for instance, may decide to move directly from his view of the current situation to policy decision, short-cutting the three stages of the process described above. Thus, to assess the soundness of the decision process up to this stage, it is useful to ask questions such as the following:

1. How was the decision made—did it go through all the stages of policy analysis?
2. How radical a departure is the decision from current policy?
3. How consistent is this decision with policies of other sectors?
4. Is the policy diffusely articulated or is it stated in a manner which is easily measurable?
5. Does the policy seem operational or is its implementation implausible?

E. Planning Policy Implementation

Once a policy has been chosen, planning for policy implementation should begin immediately. Although much of the work that must be carried out during this stage can be based on evaluations performed to make the policy decision, planning for implementation involves a concreteness absent in earlier stages of the policy process.

What was abstract during the evaluation stage begins to become concrete during planning. A schedule for moving people, physical objects and funds must be drawn up with a clarity and attention to detail that leaves no doubt as to who will do what, when and how; physical resources, once the content of hypothetical lists, must be located and their availability assured; financial resources, once ear-marked for possible use, must be appropriated so that implementation delays are minimal; the personnel needed to put plans into action must be freed from other commitments and made ready to go to work; the technical knowledge needed to guide the policy implementation must be mastered by those who will employ it; and the administrative systems within which the policy will be directed must be clearly structured and firmly in place.

Ambitious as these tasks are, there is one planning task that is more difficult (and it is the most often over-looked). This is the task of mobilizing political support. The mobilization of political support resonates most clearly when one thinks of the need to ensure that the providers and consumers of a new educational initiative embrace it with enthusiasm. Plans must be developed so that students and their families are aware of the objectives of a new initiative, that communities learn of benefits for the collectivity; programmes for teachers, educational administrators and their representatives must similarly be developed. Since new initiatives usually mean some form of job re-definition, it is important that educators see this as beneficial and that those who object to the changes be isolated. Political mobilization may also be necessary to ensure that materials for school construction are available when needed, that needed institutional administrative adjustments are carried out, and, especially, that funding proposals are approved. One important strategy for mobilizing political support is that of involving groups affected by the new initiative in the planning process. This will pay dividends not only in the form of enhanced support, but, more likely, in terms of an improved policy design.

A significant amount of planning and even de facto policy formulation take place during actual implementation. This is the case because, during implementation, the following is the rule rather than the exception:

(a) circumstances related to implementation constraints cause policy modifications to take place;

(b) feedback obtained during implementation causes reassessment of aspects of the policy decision and subsequent modifications by policymakers; and

(c) the mere translation of abstract policy intentions into concrete implementation causes re-assessment and re-design. These changes occur with great frequency because, unfortunately, implementation problems are often greatly under-estimated during the stage of policy planning.

Misjudging ease of implementation is, perhaps, the most frequent error in policy planning. No matter how deeply the various groups affected by a new initiative have been involved in reviewing and shaping plans, the concreteness of the first day of a new programme, often casts it in a new light. Implementation is the time when one discovers that schedules are unrealistic and that programmes are over-ambitious; it is the time when the ravages of inflation cause the teachers' union to demand a pay increase prior to using the new texts; it is the time when parents conclude that the certification offered by the new programme may not guarantee their children the jobs they hoped for; and it is the time when local politicians decide that they should block the initiative since it will be so successful that it will prove that the politicians in the capital are better providers than they. Such problems are often replays of issues raised during the stages of evaluating policy options or of planning, and need to be solved by taking a flexible approach to the stage of policy implementation.

No matter how well anticipated, policy implementation always brings some surprises. These shape the policy output, sometimes in crucial ways. One way to use such surprises to improve policy outcomes is to design the implementation in stages. If unanticipated problems arise at a given stage, then a re-evaluation of the plans for implementation, and possibly of the policy decision itself, is in order. Another way is to conduct well designed pilot studies, before full implementation of any projects. Problems of going to scale and the dangers of the 'greenhouse' projects that cannot survive implantation in the real world are well treated in Kemmerer (1990).

F. Policy Impact Assessment

Once the policy has been in place long enough to produce results, a policy assessment check can take place. To carry this out, it is necessary to have some sense of how long it should take for the policy, once implemented, to take hold. While policy output measurement can be carried out on a continual basis, premature attempts at assessment can mis-state the effectiveness of the policy. Furthermore, it is preferable to delay final assessment until a number of teaching cycles have transpired to separate the effect of the content of the policy change from the excitement which often accompanies implementing a new initiative for the first time. On the other hand, the sooner accurate assessment takes place, the sooner policy-makers can know if their initiatives are working as anticipated or if adjustments in policy design or policy implementation are required.

If assessment reveals that the policy outcome is lacking, it is necessary to determine whether the policy itself is inadequate, or whether poor implementation is at fault. Human capital inadequacies, under-funding, or inadequate economic stimulus during the implementation stage are among the many possible causes of failure of a well designed policy. On the other hand, if assessment reveals deficiencies in outcomes and if implementation can be shown to have been well done, then it is necessary to re-examine the policy decision and to determine what adjustments or what new policies should be substituted for the original choice. Once this is accomplished, then one moves again to the planning and implementation stages. Given the rapid pace of contemporary change and the intimate links between the educational system and the rest of society, even successfully conceived and implemented initiatives require adjustments over time.

Policy impact assessment is carried out using the same criteria employed during the policy evaluation stage. The assessment process revolves around the following questions: What have been the actual impacts of the policies in question? Are these impacts desirable given the changes that were hoped for? Are the changes affordable? Did costs prevent their full implementation? Did cost over-runs make it unthinkable to implement them over a longer term or on a wider basis? Can the policy be lived with politically and socially? Are the impacts feasible? Were full impacts accomplished? Would exceptional efforts be required to replicate these impacts in other circumstances?

G. Subsequent Policy Cycles

If a policy initiative is carried out systematically, the process of policy design, planning, implementation, impact assessment, and re-design will become iterative, and, in theory, infinitely so, as Figure 13.3 suggests. Unfortunately, long-term policy analysis and planning is not often carried out in such a fashion. Often the results of verification are not ploughed back into policy. Instead, verification is often seen as a stock-taking exercise, needed in order to close the books on a policy initiative. Later in the country's history, when policy change is once again needed in the educational area under discussion, a policy process often begins de novo and may duplicate much of the analysis, derivation of alternative options, evaluation, and planning carried out earlier. The conclusion, then, of policy analysis is never to conclude. Ideally, once implementation has been completed and policy outcomes are forthcoming, a policy impact assessment stage ensues, leading potentially to a new policy cycle.

REFERENCES

Allison, G. T. 1971. Essence of decision. Boston: Little, Brown and Company.

Carley, M. 1980. Rational techniques in policy analysis. London: Heinemann Educational Books.

Coombs, P.; Hallak, J. 1987. Cost analysis in education: a tool for policy and planning. Baltimore, Maryland: World Bank and Johns Hopkins University Press.
Haddad, W.D.; Demsky T. 1994. The dynamics of education policy making. Washington D.C.: The World Bank.
Kemmerer, F. 1990. "Going to scale: why successful instructional development projects fail to be adopted", in D. Chapman and C. Carrier (Eds.), Improving educational quality: a global perspective. Westport, Connecticut: Greenwood Press, 243-256.
Kemmerer, F. 1994. Utilizing education and human resource sector analyses. Fundamentals of Educational Planning series, No.47, Paris, UNESCO: International Institute for Educational Planning.
Lindblom, C.; Cohen D. K. 1979. Usable knowledge: social science and social problem solving. New Haven: Yale University.
Mingat, A.; Jee-Peng Tan. 1988. Analytical tools for sector work in education. Baltimore, Maryland: World Bank and Johns Hopkins University Press.
North, D. C. 1990. Institutions, institutional change and economic performance. Cambridge; New York: Cambridge University Press.
North, D.C. 1994. "The new institutional economics and development", Forum 1 (2) May, 1994: 3-6.
Windham, D. 1988a. "Effectiveness indicators in the economic analysis of educational activities", International Journal of educational research, 12(6): 575-666.
Windham, D. 1988b. Indicators of educational effectiveness and efficiency. Tallahassee, Florida: IEES project.

Reading 13.2

NEA GUIDE TO EDUCATOR-LED SCHOOL IMPROVEMENT AND THE GPS INDICATORS FRAMEWORK

THE PATH TO SUCCESS

Student success. These two words define the career and aspirations of three million professional educators. The words are simple, but the task is not. Students come into America's public school classrooms with very different needs, backgrounds, levels of preparation, and support. Educators work hard to develop the potential, independence, and character of every student who shows up in our classrooms. But we know there are very real barriers to students' academic success, including the underlying issues of hunger, poverty, and prejudice that has concretized systemic racism and privilege, family instability, and illness. We desperately want to help students overcome these obstacles to achievement. We continually ask ourselves how we can elevate all of our students and give them the knowledge and

skills they need to thrive in this challenging 21st century world. We are committed to creating great public schools for all students. That's why in 2008, the National Education Association (NEA), in collaboration with the nation's top education experts, developed the Great Public Schools (GPS) Indicators Framework (nea.org/gpsindicators). The GPS Indicators Framework showcases the research- and evidence-based resources, policies, and practices that are integral to making our public schools great—these seven criteria are crucial to student success.
The seven criteria are:

- Quality programs and services that meet the full range of all children's needs so that they come to school every day ready and able to learn
- High expectations and standards with a rigorous and comprehensive curriculum for all students
- Quality conditions for teaching and lifelong learning
- A qualified, caring, diverse, and stable workforce
- Shared responsibility for appropriate school accountability by stakeholders at all levels
- Parental, family, and community involvement and engagement
- Adequate, equitable, and sustainable funding

We believe these criteria will:

- Prepare all students for the future with 21st century skills
- Create enthusiasm for learning and engage all students in the classroom
- Close achievement gaps and raise achievement for all students
- Ensure that all educators have the resources and tools they need to get the job done

HOW TO USE THIS GUIDE

It is not enough to have a framework; we must actually use this framework to create real change within our schools. This guide was designed for leaders striving for excellence in their schools and aims to assist educators in their efforts to transform their schools into great public schools and demonstrates how to apply the GPS Indicators Framework in the school improvement process. The guide also shows how the Framework informs other strategic planning tools such as the SWOT grid and Logic Model. Further, although not the focus here, the guide can be used to inform and develop a student-centered campaign. The school improvement process closely resembles the student-centered campaign process.

The Framework not only addresses the research- and evidence-based resources, policies, and practices proven to narrow opportunity and skills gaps, but also enables users to evaluate how well states, districts, and schools provide and implement the resources, policies, and practices critical to student success. In other words, the indicators within the Framework can be used to pinpoint gaps between what currently exists within your state, district, and/or school and what is needed to create a great public school.

The transformation into a great public school will require work—specifically teamwork—stakeholder involvement, dedication to change, and several other tools (identified later in this guide) in addition to the GPS Framework. After the process, however, users will have a clear set of priorities, targets, and a plan for their schools.

This guide is an abridged version of the school improvement process and focuses on the assessment and planning phases, only.

ASSESS—IDENTIFY THE PROBLEM(S)

After completing the first phase of the improvement process, use the GPS Indicators Framework to begin identifying gaps between your school and a great public school. You can find a copy of the Framework at nea.org/gpsindicators.

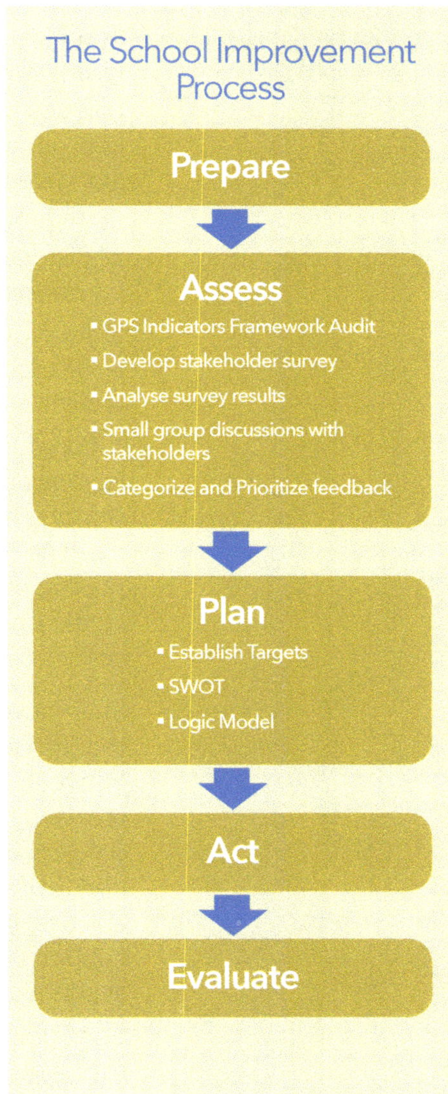

1. GPS Indicators Framework—Initial Audit

Have your team draft a list of challenges. Initiate the conversation by asking guiding questions:

- What is working in our district/school?
- What are the gaps in services?
- Where are there duplications?
- What services should be linked?
- Are services accessible?

After the brainstorm, go through the entire GPS Framework with your team. The Framework functions as an auditing tool, assisting users in identifying gaps between current school resources, policies, and practices and necessary resources, policies, and practices.

GPS Indicators Framework Sample:

CRITERIA	**STANDARDS AND CURRICULUM**			**ABC HIGH SCHOOL AUDIT**
SUBCRITERIA	**COMPREHENSIVE CURRICULUM CONTENT**			
INDICATORS	Resources	State provides funding to implement rigorous courses aligned with college- and career-ready standards for all districts.*	X	• Our school has received funding for two Honors English courses, one AP English course, and one AP history course, only. • Our school does not have any fine arts courses. • Our school no longer provides physical education (P.E.).
		State provides funding to all districts for fine arts education.	X	
		State provides funding to all districts for physical education.	X	
		** Rigorous courses could include dual enrollment, Honors, Advanced Placement (AP), International Baccalaureate (IB), and career and technical education (CTE) certification*		
INDICATORS	Policies & Practices	State developed a policy that requires alignment between curricular content and rigorous standards that address the needs of students of all abilities, linguistic, and cultural backgrounds in all academic subjects.	X	• Our school has rigorous standards for all students, but only has AP English and history courses. • The state has no such policy. Our school has no policy stating the importance of fine arts.
		State policy recognizes the value of fine arts in curricula.	X	

CRITERIA	STANDARDS AND CURRICULUM		ABC HIGH SCHOOL AUDIT
SUBCRITERIA	COMPREHENSIVE CURRICULUM CONTENT		
	State policy recognizes physical education as a core subject.	X	• The state recognizes P.E. as a core subject, but our school has no such policy.
	Schools align curriculum content to rigorous standards that address the needs of students of all abilities, linguistic, and cultural backgrounds in all academic subjects.	X	• Our school has rigorous standards for all students, but only has AP English and history courses.
	Schools offer fine arts education to their students.	X	• Our school does not offer fine arts.
	Schools implement the National Association of Sport and Physical Education (NASPE) standards for physical education.*	X	• Our school no longer offers P.E.
	Schools use the community as a contextualized learning environment.**	X	• Our school does provide field trips but does not offer service learning, and we no longer have a school library.
	**NASPE recommends 150 minutes of instructional physical education for elementary school students and 225 minutes for middle and high school students per week for the entire school year.*		
	***Connect education to community through public libraries, zoos, parks, work experience opportunities, service learning, the school library, and after-school programs.*		

CRITERIA	STANDARDS AND CURRICULUM			ABC HIGH SCHOOL AUDIT
SUBCRITERIA	COMPREHENSIVE CURRICULUM CONTENT			
INDICATORS	Outputs	Percentage of students enrolled in a Gifted and Talented education program.	X	• 0 percent–Our school does not have this program.
		Percentage of students enrolled in at least one AP course.	X	• 60 percent of our students are enrolled in at least one AP course.
		Percentage of high school seniors who have completed all college-preparatory courses in math and science.*	X	• 0 percent–Our school does not offer physics or calculus.
		Percentage of students enrolled in a fine arts course.	X	• 0 percent–Our school does not offer fine arts.
		Percentage of students enrolled in a P.E. course that meets NASPE standards.	X	• 0 percent–Our school does not offer P.E.
		Percentage of students participating in service learning and/or an afterschool program.	X	• 70 percent of our students participate in afterschool programs.
		**College preparatory courses are algebra I, algebra II, geometry, trigonometry, calculus, biology, chemistry, and physics.*		

ABC High School Example

It is clear from the audit that the school does not have the necessary resources, policies, or practices to meet satisfactory output levels. A great public school would have all of the required inputs and would produce satisfactory outputs. For this particular subcriterion, a great public school would have funding for a wide range of courses, would implement courses like P.E., fine arts, and calculus, and would have a high percentage of students completing college preparatory courses. The team reviews the audit and agrees the lack of rigorous courses (i.e., Honors and AP) and college preparatory courses (e.g., calculus and physics) are major issues.

2. Survey

After identifying gaps with your team, develop a survey for other stakeholders in the school improvement process. Surveys are a great way to solicit feedback from other staff, parents, students, and community members. The survey will either confirm challenges and priorities already discovered by your team, highlight challenges and priorities previously unknown by your team, or both. Survey tips:

- Administer the survey at existing meetings such as PTA, community meetings, and staff meetings.
- Insert questions based on the GPS Indicators Framework audit.
- Surveys should not exceed 25 questions.
- Surveys should be designed to force respondents to make a choice.
- Aim for a 95 percent confidence level with a +/–2 or 3 sampling error.
- Surveys should represent all stakeholders (e.g., parents, students, community members, school staff) and may take a few months to collect.

The following are modified sample survey questions from Robert Ewy's guide on strategic planning:[1]

1. What are the greatest challenges our school faces over the next five years (choose only five of your highest priorities from the list)?
   - Preparing students for college
   - Hiring and retaining quality educators
   - Providing additional supports to at-risk students
   - Meeting standardized testing benchmarks
   - Meeting the needs of students with disabilities
   - Maintaining sports programs
   - School climate
2. What should the financial priorities be over the next five years (choose only five of your highest priorities from the list)?
   - Building maintenance
   - Updated technology in classrooms
   - Specialized Instructional Support Personnel (e.g, nurses, psychologists, social workers, counselors) to meet student needs

- The addition of rigorous courses (e.g, Advanced Placement courses)
- Small class sizes
- New textbooks
- Teacher salaries and benefits to improve recruitment and retention
- A school library
- College preparatory courses for all students
- School climate

3 What evidence do you use to evaluate the quality of education in our school (choose only five of your highest priorities from the list)?

- Presence of a school nurse
- Fine arts courses
- P.E. course
- Advanced courses in math and science (e.g., calculus and physics)
- Standardized test scores
- School climate survey results
- Teacher turnover
- Student absences
- Percent of teachers teaching out of field
- Parent engagement
- Percentage of students being suspended

For additional information on survey development contact NEA's Research Department at insights@nea.org.

[1] *Ewy, R. (2009). Stakeholder-Driven Strategic Planning in Education: A Practical Guide for Developing and Deploying Successful Long-Range Plans. Milwaukee, Wisconsin: ASQ Quality Press.*

3. Analyze Survey Results

ABC High School Example

During the GPS Indicators audit, your team identified a lack of rigorous courses (i.e., Honors and AP courses) and college preparatory courses as a priority. Surveys and small group discussions revealed college preparatory courses and school climate as school priorities. Due to the team's limited capacity, the team decides to focus on college preparatory courses, only.

After collecting all of the surveys and any data collected from small-group discussions, analyze and prioritize the results. Results will be used to develop the first draft of your school improvement plan.

PLAN—IDENTIFY THE SOLUTION(S)

The action plan is based on the prioritized data collected in the assessment phase.

1. Establish Targets

After setting priorities, the team should establish targets. This section enables the team to establish where they currently are in regard to the priority and where they would like to be. Target-setting will later inform the SWOT and logic model.

ABC High School Example

College preparatory courses were voted as a top priority. College preparatory courses are algebra I, algebra II, geometry, trigonometry, calculus, biology, chemistry, and physics.

Targets Sample

PRIORITIES	CURRENT STATUS	TARGETS
Percentage of high school seniors who have completed all college preparatory courses in math and science	0 percent of students have completed the complete course catalog of college prep courses	85 percent completion rate after five years
	60 percent of seniors completed algebra I, algebra II, geometry, and trigonometry	5 percent annual increase in the number of students completing math preparatory courses
	60 percent of seniors completed biology and chemistry	5 percent annual increase in the number of students completing science preparatory courses
	0 calculus and physics courses	Hire a teacher for calculus and a teacher for physics

2. SWOT

The SWOT (Strengths, Weaknesses, Opportunities, and Threats) analysis helps identify possible challenges your action plan may face and will inform strategy development. Information and data from your GPS audit, survey, and discussions will assist your team in filling out your SWOT grid.

Strengths-Characteristics that will help you achieve your objectives
Weaknesses-Characteristics that will limit your ability to achieve objectives
Opportunities-External elements that can be seized to achieve objectives
Threats-External elements that negatively impact your ability to achieve objectives

-SAMPLE- SWOT ANALYSIS		
	HELPFUL	HARMFUL
INTERNAL	**Strengths** • The school set rigorous standards for all students • Educators meet high-quality standards • Strong parent engagement	**Weaknesses** • Course offerings are not aligned with high student standards • Average rating on the school climate student surveys • Inactive student council • 40 percent of the school is not completing any college prep courses
EXTERNAL	**Opportunities** • An active PTA • Well-attended school board meetings	**Threats** • Funding constraints • Declining school enrollment

TOWS Analysis

The TOWS (Threats, Opportunities, Weaknesses, and Strengths) analysis is a variation of the SWOT analysis, but with more emphasis on the external environment. For some, the TOWS helps clarify their options.

Strengths and Opportunities (S-O) Strategies-How strengths can be used to capitalize on opportunities
Strengths and Threats (S-T) Strategies-How strengths can be used to mitigate threats
Weaknesses and Opportunities (W-O) Strategies-How opportunities can overcome weaknesses
Weaknesses and Threats (W-T) Strategies-How you will minimize weakness and avoid threats

INTERNAL / EXTERNAL	STRENGTHS	WEAKNESSES
	• The school set rigorous standards for all students • Educators meet high-quality standards • Strong parent engagement	• Course offerings are not aligned with high student standards • Average rating on the school climate student surveys • Inactive student council • 40 percent of the school is not completing any college prep courses
OPPORTUNITIES	**S-O STRATEGIES**	**W-O STRATEGIES**
• An active PTA • Well-attended school board meetings	• Have educators inform parents of standards • Have educators work with the board and PTA, reaffirming high standards for all students	• Work with the PTA and the board to increase enrollment in current courses and achieve additional courses • Demonstrate to the board and PTA that misalignment between courses and standards has negatively impacted school performance and school climate
THREATS	**S-T THREATS STRATEGIES**	**W-T STRATEGIES**
• Funding constraints • Declining school enrollment	• Demonstrate that high standards require college prep courses, all of which require funding • Demonstrate that high standards and college prep courses can attract parents and positively impact school enrollment	• Implement courses aligned with standards • Implementation of college prep courses may positively impact student morale • A campaign to implement college prep courses could activate the student council • A strong curriculum could positively impact enrollment

3. Logic Model

The logic model enables your team to visualize the resources and activities needed to address priorities and achieve established targets. The logic model will serve as the foundation for your more detailed action plan. There are many variations of the logic model. The example below is one of many ways to create your logic model. A separate logic model must be created for each objective identified.

Priority 1, Objective A: Increase enrollment in current college preparatory courses

					OUTCOMES		
PRIORITIES	**TARGETS**	**RESOURCES**	**ACTIVITIES**	**OUTPUTS**	**SHORT-TERM OUTCOMES**	**MEDIUM-TERM OUTCOMES**	**LONG-TERM OUTCOMES**
		Resources needed to meet our objectives:	*To address our problem or asset, we will accomplish the following activities:*	*We expect that once accomplished, these activities will produce the following evidence of service delivery:*	*We expect that if completed or ongoing, these activities will lead to the following changes in one year:*	*We expect that if completed or ongoing, these activities will lead to the following changes in two years:*	*We expect that if completed, these activities will lead to the following changes in three years:*
Percentage of high school seniors who have completed all college preparatory courses in math and science	85 percent completion rate after five years	Staff time Parent and stakeholder time Printing funds for informational leaflets and pledge cards	Research and survey students not currently enrolled in college preparatory courses	Attendance rates at meetings Number of administrators supportive of increased enrollment	June 2016 65 percent of seniors completed all available college preparatory courses	June 2017 70 percent of seniors completed all available college preparatory courses	June 2018 75 percent of seniors completed all available college preparatory courses

					OUTCOMES		
PRIORITIES	TARGETS	RESOURCES	ACTIVITIES	OUTPUTS	SHORT-TERM OUTCOMES	MEDIUM-TERM OUTCOMES	LONG-TERM OUTCOMES
	5 percent annual increase in the number of students completing math preparatory courses 5 percent annual increase in the number of students completing science preparatory courses		Meet with parents, community, and students about the importance of taking college preparatory courses Meet with administration and staff on plan to increase enrollment numbers Monthly meetings with student council on enrollment	Number of pledge cards signed by students Enrollment rates in courses			

(*continued*)

Priority 1, Objective B: Acquire additional college preparatory courses

					OUTCOMES		
PRIORITIES	TARGETS	RESOURCES	ACTIVITIES	OUTPUTS	SHORT-TERM OUTCOMES	MEDIUM-TERM OUTCOMES	LONG-TERM OUTCOMES
		Resources needed to meet our objectives	*To address our problem or asset, we will accomplish the following activities:*	*We expect that once accomplished, these activities will produce the following evidence of service delivery:*	*We expect that if completed or ongoing, these activities will lead to the following changes in one year:*	*We expect that if completed or ongoing, these activities will lead to the following changes in two years:*	*We expect that if completed, these activities will lead to the following changes in three years:*
Percentage of high school seniors who have completed all college preparatory courses in math and science	85 percent completion rate after five years 5 percent annual increase in the number of students completing math preparatory courses	Staff time Parent and stakeholder time Printing funds for informational leaflets and pledge cards	Research for presentation to the board and administrators Meet with parents, community, and students about acquiring additional college prep courses	Attendance rate at monthly meetings Number of pledge cards signed by all stakeholders Number of administrators supportive of additional courses	June 2019 Additional college prep courses are added to the school catalog	June 2020 80 percent of seniors completed all college preparatory courses	June 2021 85 percent of seniors completed all college preparatory courses

					OUTCOMES		
PRIORITIES	TARGETS	RESOURCES	ACTIVITIES	OUTPUTS	SHORT-TERM OUTCOMES	MEDIUM-TERM OUTCOMES	LONG-TERM OUTCOMES
	5 percent annual increase in the number of students completing science preparatory courses		Meet with administrators Schedule stakeholder gatherings at the school board meetings and address the board Prepare stakeholders to address the board Monthly informational meetings with updates on the campaign to achieve college prep courses	Attendance at the presentation for the board Number of supportive stakeholders at board meetings			

4. Detailed Action Plan

At this point, your team should have completed the following:

- Identified gaps between your school and a great public school
- Met with and surveyed stakeholders on school priorities and challenges
- Categorized and prioritized the feedback received from the audit, survey, and small group discussions
- Established targets for identified school priorities
- Identified strengths, weaknesses, opportunities, and threats (SWOT)
- Developed a logic model based on targets, priorities, and the SWOT

Now that you have a solid foundation, write up your final plan. The final plan should include detailed timelines, milestones, and assigned duties.

Sample Action Plan

GOAL: COLLEGE-READY STUDENTS						
OBJECTIVES	**ACTIVITIES**	**TIMELINE**	**ROLES**	**STRATEGIES**	**MEASURES**	**RESOURCES NEEDED**
Increase enrollment in current college preparatory courses	Meet with administration and staff on plan to increase enrollment numbers Research and survey students not currently enrolled in college preparatory courses Meet with parents, community, and students about the importance of taking college preparatory courses Meet monthly with student council on enrollment	9/2015 10/2015 2/2016–12/2022 8/2016–12/2022	Mary–Supervisor Brenda–Research and surveys Tom–Logistics Kate–Liaison to parents Marcus–Liaison to community members Bruce–Liaison to administration Kelly–Liaison to student council Cory–Liaison to staff	Surveys Provide information via meetings Solicit feedback via meetings	Number of administrators voicing support Number of staff voicing support Number of meetings held Survey response rate Meeting attendance Number of pledge cards signed by students Enrollment rates in courses	Assistance from school staff Assistance from stakeholders $100 for leaflets $200 for pledge cards $40 for refreshments for each meeting with parents

GOAL: COLLEGE-READY STUDENTS						
OBJECTIVES	**ACTIVITIES**	**TIMELINE**	**ROLES**	**STRATEGIES**	**MEASURES**	**RESOURCES NEEDED**
Acquire additional college preparatory courses	Meet with administrators about acquiring additional courses Meet with parents, community, and students about acquiring additional college prep courses Research for presentation to the board and administrators Prepare stakeholders to address the board Schedule stakeholder gatherings at the school board meetings and address the board Monthly informational meetings with updates on the campaign to achieve college prep courses	1/2017 2/2017–5/2020 8/2017 9/2017–5/2020 10/2017–6/2020 11/2017–7/2020	Mary–Supervisor Brenda–Research and surveys Tom–Logistics Kate–Liaison to parents Marcus–Liaison to community members Bruce–Liaison to administration Kelly–Liaison to student council Cory–Liaison to staff	Provide information via presentations and meetings Solicit information via feedback	Number of administrators and staff supportive of additional courses Number of meetings Attendance rate at monthly meetings Number of pledge cards signed by all stakeholders Attendance at the board meetings Number of supportive stakeholders at the board meeting Number of courses implemented	Assistance from school staff Assistance from stakeholders $100 for leaflets $200 for pledge cards $40 for refreshments for each meeting with parents

A FINAL WORD

Transforming your school will not be easy; it will require a dedicated team, resources, and a commitment to work with all stakeholders at every step. Depending on your particular school and the priorities identified by all stakeholders, including your team, the school transformation process can take anywhere from two to 10 years. Although arduous, the process will change your school into a great public school and provide the high-quality education all students deserve.

Let your school be the example—the lighthouse that guides other schools toward greatness. Only when all stakeholders work together, can we eliminate inequities, close the skills gap, and elevate achievement for all students.

For more information, please contact Angelica Castañon in NEA's Education Policy and Practice Department at acastanon@nea.org.

NOTES

HOW DO POLICY CONSIDERATIONS AND IMPLICATIONS APPLY TO MY PROFESSION?

All educators are affected by the policies that are adopted in our educational systems. As indicated in the literature wherein, there are multiple considerations and implications of each adopted policy. So, it is important that educators remain vigilant yet flexible. One of the best ways to have a voice in decision-making processes is to become the expert in your field. The more knowledge you obtain, the more presence you may have in various constituent groups and with educational stakeholders. According to the National Education Association (NEA, n.d.), students, parents, community leaders, and educators alike have a responsibility to advocate for advancing educational success. Some specific examples include teachers working with educational leaders to establish collaborative teams, leaders can work with the community to establish partnerships that increase voices, and students along with their families can demand equitable funding and support services. When all stakeholders work together for the common good of the educational landscape, inequities can be eliminated and achievement for all students can be increased.

GLOSSARY OF TERMS

Ability-referenced interpretations: An interpretation based on the student's test performance compared to what we believe the student should be able to do based on his/her ability.

Accountability: Taking responsibility or liability for actions, decisions, or policies.

Achievement tests: Tests used to measure knowledge and skills.

Ambiguous temporal precedence: Uncertainty about whether the cause precedes the effect.

Aptitude tests: Tests used to assess an individual's potential or future performance (e.g., predict future job performance).

Assessment: A general term that describes the many techniques we use to measure and judge student behavior, ability, skills, and performance.

Attrition: A threat to both internal and external validity that refers to the failure of participants to complete all outcome measures.

Central tendency: A statistic describing the center of a dataset of observed scores as measured by the mean, median, and mode.

Classical test theory (CTT): A theory that postulates that measurement error, a random latent variable, is always part of an observed score.

Concurrent validity: The extent to which the test scores estimate individuals' present standing on the criteria related to an already validated measure.

Confidence intervals (CIs): A range or band of scores that have a specified probability of including the true score.

Compensatory rivalry: The idea that participants of the control groups can feel competitive to do as well as the treatment group, despite not getting treatment benefits.

Construct-irrelevant test variance: The test variance indicates the test includes excess reliable variance that is unnecessary for interpretation of the construct, such as construct-irrelevant difficulty (i.e., difficult language) or construct-irrelevant easiness (i.e., too much time on a test).

Construct validity: The degree to which certain theoretical or explanatory concepts or constructs account for performance on the test.
Content validity: The degree to which a test measures all facets of the construct it is designed to measure.
Convergent evidence: Signifies the measure in question is coherently related to other measures of the same construct based on theoretical grounds.
Criterion-referenced interpretations: Test scores that are compared to some extant standards or criteria.
Criterion-related validity: The extent to which test scores relate to one or more external variables (called criteria), which are considered to provide a direct measure of the characteristic or behavior in question.
Diagnostic assessment: Any type of assessment used to identify students' knowledge or skills prior to instruction.
Dichotomous response tests: Tests that include items with two possible answer choices (e.g., right/wrong, yes/no, true/false).
Discriminant evidence: Signifies the measure is not related to exemplars of other distinct theoretical constructs. This is critical for discounting other plausible hypotheses.
Evaluation: Involves the use of measurement to make decisions or judgments about or values of something.
External reliability: Refers to the extent to which a measure varies from one use to another (i.e., test re-test).
External validity: The generalizability of the inferences; that is, the extent to which the conclusions are true for variations in persons, settings, and times.
Face validity: The extent to which a test appears to measure what it was intended to measure.
Formative assessments: Assessments used while an instructional unit is in progress. These are used primarily to give the teacher feedback on how the students are progressing.
Growth-referenced interpretations: An interpretation based on comparing the student's test score after instruction with the score of a similar test given prior to instruction.
History: The events that occur between the beginning of the treatment and the posttest that could have produced the observed outcomes in the absence of that treatment.
Instrumentation: Term used to describe the tools used to measure various qualities and quantities.
Intelligence tests: Instruments that measure a latent construct, typically a group of related variables (e.g., verbal, comprehension, memory), that we call "intelligence" or "IQ," which refers to an intelligence quotient.

Interest inventories: Tests used to measure at what an individual is best or for what he/she is best suited (e.g., career choice).

Internal consistency reliability: The extent to which test items designed to measure the same concept or construct actually do so.

Internal validity: The ability to make causal inferences among the relationship between two variables (X and Y).

Inter-rater or Inter-observer reliability: The degree to which different raters or observers give consistent estimates of the same phenomenon.

Interval: The name given to data that are continuous and that have equal distances between points.

Item analysis: A post hoc analysis performed to compare the performance on each item of the most and least successful examinees on the total test.

Item difficulty: The proportion of students who answered the item correctly.

Item discrimination index: Refers to the extent to which an item is able to distinguish between high and low scorers on the test.

Kurtosis: Refers to the height of the peak of the data.

Latent construct: A variable that is not directly measurable or observable.

Maturation: The idea that participants in a study will experience natural changes that would occur even in the absence of treatment.

Maximum performance measures: Measures used when we want to assess how well our students can perform.

Median: The 50th percentile. This is the point that divides the distribution of scores into two equal halves.

Mean: Often thought of as the average of the scores; calculate the mean by adding all scores and then dividing by the total number of scores.

Measurement: The process of assigning meaningful numbers (or labels) to persons or objects based on the degree to which they possess some characteristic.

Measures of dispersion: The extent to which scores in a data set are spread out or differ from each other. Common examples include variance, standard deviation, and range.

Measures of relative position: A value or score that represents the distance from a reference point. Common examples include percentile scores and z-scores.

Measurement variance: A measure of the spread between numbers in a data set.

Mode: The score that appears most frequently in the data set.

Multiple response tests: Tests that include a stem and leaves; commonly known as multiple-choice tests.

Nominal: The name given to categorical data used simply as an identifier.

Normal distribution: A distribution of scores that is perfectly symmetrical. It has both a skewness and kurtosis of zero.

Norm-referenced interpretations: Interpretations based on comparing the score of one student with the scores from some norm group.

Novelty: The idea that the introduction of an intervention can breed excitement, energy, and so on that can increase the possibility of success.

Observed scores: The number or percentage correct that the student receives.

Observed variables: Objects that are directly measurable using scales.

Ordinal: The name given to data that represent a series of relationships or a rank. The distance between each data point is not the same. Common examples of ordinal data are Likert scales such as Strongly Agree, Moderately Agree, Agree, Somewhat Disagree, Strongly Agree.

Outlier: A data point that lies an abnormal distance from other data points in a dataset.

Parallel-forms reliability: Reliability estimate calculated by the correlation between two versions of a test that are designed to measure the same content.

Percentiles: The term used to describe the relative position of a score or the percentage of scores that fall below a particular score.

Performance assessment: Refers to the measuring of the demonstration of a skill.

Personality tests: The measure of traits and characteristics of an individual (e.g., attitudes, values).

Pivot table: A chart used to summarize, sort, and group quantitative data to illustrate various categories.

Predictive validity: Indicates the extent to which individuals' future performance on the criterion (outcome) is predicted from prior test performance.

Preliminary or placement assessments: Assessments designed to measure students' prerequisite skills.

Quartile: A measure of relative position by which data are divided into four equal groups.

Random error: Sometimes called internal error; refers to effects that impact the results of the particular assessment for specific individuals at that time.

Range: The difference between the highest and the lowest scores in the distribution.

Ratio: The name given to data that has an absolute zero and no negative values (i.e., height).

Raw score: The number of items correctly answered on a given test.

Reactivity: The idea that the factor structure and level of responses can be affected by whether a person is or is not accepted into the treatment group even before that person receives treatment.

Regression toward the mean: A statistical tendency or phenomenon that occurs when a score is extreme on the first measurement, it will tend to be closer to the mean on future measurements. .

Reliability: The degree to which the results of a measurement are stable, consistent, and accurate when they are repeated under the same conditions. .

Reliability coefficient: The measure of accuracy of scores typically obtained by calculating the correlation between at least two different instruments, the same instrument administered twice, or related items on a test.

Research: The study, investigation, or examination of one or more variables to discover or interpret facts.

Resentful demoralization: The idea that members of a group receiving a less desirable treatment, or no treatment, can be resentful, which can change their responses to the outcome measures.

Scaled scores: The result obtained when a statistical transformation is applied to the raw scores to make scores across examinees consistent over time.

Selection: The systematic differences between the average person in one condition relative to another.

Skewed distribution: When the tail of the distribution points to the left or the right.

Split-half reliability: Estimate of the extent to which particular items comprising two forms of a single test seem to be measuring the same content domain.

Standard deviation (SD): Describes the spread of scores around the mean.

Standard error of measurement (SEM): An estimate of how repeated measures of the same student on the same test are distributed around that student's true score.

Summative assessments: Assessments performed at the end of instruction to determine the students' level of competency with or understanding of the material. Results are often included in grade calculations.

Systematic error: Errors that impact the entire sample or group of test takers and that are consistent across administrations of the test.

Testing (threat to internal validity): A threat to internal validity in pretest/ posttest scenarios which occurs when the experience taking the initial test influences the scores on a retest.

Test-retest reliability: Estimate of the relationship of scores when two administrations of the the same form of the test have been completed.

True score: The actual score from a measure (knowledge, skills, attitude, height, etc.) with the absence of error

Typical performance measures: Measures designed to assess students' most common, or typical, behavior or performance.

Underrepresentation: Occurs when a test does not include a sufficient number of items to measure all facets of a construct. Underrepresentation can lead to invalid interpretations.

Validity: The truth of, correctness of, or degree of support for an inference.

Variance: The square of the standard deviation. The term used to describe the average squared difference of scores from the mean score of a distribution.

***z*-score**: A measure of how many standard deviations the raw score is from the population mean. Z-scores are useful measures of relative position and are often called standard scores.

REFERENCES

Bill & Melinda Gates Foundation. (2010). *MET project: Working with teachers to develop fair and reliable measures of effective teaching.* Retrieved from http://k12education.gatesfoundation.org/download/?Num=2813&filename=met-framing-paper.pdf

Campbell, D. T., & Fiske, D. W. (1959). Convergent and discriminant validation by the multitrait-multimethod matrix. *Psychological Bulletin, 56*(2).

Campbell, D. T., & Stanley, J. C. (1963). *Experimental and quasi-experimental designs for research*. Chicago: Rand McNally.

Common Core State Standards Initiative (2020). *Preparing America's students for success*. http://www.corestandards.org/

Cook, D. A., & Beckman, T. J. (2006). Current concepts in validity and reliability for psychometric instruments: Theory and application. *The American Journal of Medicine, 119*(2), 166–67.

Corcoran, S. P. (2010). *Can teachers be evaluated by their students' test scores? Should they be? The use of value-added measures of teacher effectiveness in policy and practice*. Education Policy for Action Series: Education Challenges Facing New York City, Annenberg Institute for School Reform at Brown University. http://www.annenberginstitute.org/sites/default/files/product/211/files/valueAddedReport.pdf

Cross, L. H., & Frary, R. B. (1999). Hodgepodge grading: Endorsed by students and teachers alike. *Applied Measurement in Education, 12*(1), 53–72.

Elementary and Secondary Education Act of 1965 (P.L. 89–10). (1965). http://www.scribd.com/doc/49149656/Elementary-and-Secondary-Education-Act-of-1965

Every Student Succeeds Act (ESSA). (2015). https://www.ed.gov/essa

Field, A. (2009). *Discovering statistics using SPSS.* London: SAGE Publications.

Jönsson, A., & Prins, F. (2019). Editorial: Transparency in assessment—exploring the influence of explicit assessment criteria. *Frontiers in Education*, 3. https://doi.org/10.3389/feduc.2018.00119

Jonsson, A. (2014). Rubrics as a way of providing transparency in assessment. *Assessment & Evaluation in Higher Education 39*(7), 840–852. https://doi.org/10.1080/02602938.2013.875117

Haddad, W. D., & Demsky, T. (1995). Education policy-planning process: An applied framework. *IIEP Studies, 51.* http://www.unesco.org/education/pdf/11_200.pdf

Harlen, W. (n.d.). On the relationship between assessment for formative and summative purposes. *Assessment and Learning*, 87–102. https://doi.org/10.4135/9781446250808.n6

Hill, M. F., Tan, K., & Jiang, H. (2018). Classroom assessment for teacher learning and student learning. *Teacher Learning with Classroom Assessment*, 193–201. https://doi.org/10.1007/978-981-10-9053-0_12

Hong, B., & Ehrensberger, W. (2007). Assessing the mathematical skills of students with disabilities. *Preventing School Failure, 52*(1), 41–47.

Kelly, T. L. (1939). The selection of upper and lower groups for the validation of test items. *Journal of Educational Psychology, 30*, 17–24.

Linn, R. L., & Miller, D. (2005). *Measurement and assessment in teaching* (9th ed.). Upper Saddle River, NJ: Prentice Hall.

Marsh, H. W., Nagengast, B., Fletcher, J., & Televantou, I. (2011). Assessing educational effectiveness: Policy implications from diverse areas of research. *Fiscal Studies, 32*, 279–295.

Messick, S. (1980). Test validity and the ethics of assessment. *American Psychologist, 35*, 1012–1027.

Messick, S. (1994). *Validity of psychological assessment: Validation of inferences from persons' responses and performances as scientific inquiry into score meaning.* Research Report. Educational Testing Services, Princeton, N.J. https://files.eric.ed.gov/fulltext/ED380496.pdf

Messick, S. (1989). Meaning and values in test validation: The science and ethics of assessment. *Educational Researcher*, 18(2), 5–11.

Moss, P. A. (1992). Shifting conceptions of validity in educational measurement: Implications for performance assessment. *Review of Educational Research, 62*(3), 229–258. https://doi.org/10.2307/1170738

National Center for Higher Education Management Systems (NCHEM). (2008). http://higheredinfo.org/dbrowser/index.php?submeasure=119&year=2008&level=nation&mode=data&state=0

National Commission on Excellence in Education (NCEE). (1983). *A nation at risk: The imperative for educational reform*. http://datacenter.spps.org/uploads/SOTW_A_Nation_at_Risk_1983.pdf

National Council of Teachers of Mathematics (NCTM). (1995). *Assessment standards for school mathematics*. Reston, VA: NCTM.

National Council of Teachers of Mathematics. (2005). *Assessing to learn and learning to assess*. https://www.nctm.org/News-and-Calendar/Messages-from-the-President/Archive/Cathy-Seeley/Assessing-to-Learn-and-Learning-to-Assess/

National Education Association (NEA). (n.d.). *NEA guide to educator-led school improvement and the GPS indicators framework*. http://www.nea.org/assets/docs/18874-Eductr-Led_Schl_Imprvmnt_TK_OL.pdf

No Child Left Behind Act of 2001 (P.L. 107–110). (2002). http://www2.ed.gov/policy/elsec/leg/esea02/107-110.pdf

Pajares, M. F. (1992). Teachers' beliefs and educational research: Cleaning up a messy construct. *Review of Educational Research, 62*(3), 307–332.

Picciano, A. G. (2012). The evolution of big data and learning analytics in American higher education. *Journal of Asynchronous Learning Networks, 16*(3), 9–20.

Randall, J., & Engelhard, G. (2010). Examining the grading practices of teachers. *Teaching and Teacher Education: An International Journal of Research and Studies, 26*(7), 1372–1380.

Ridhwan, M. (2017). Understanding formative and summative assessment for EFL teachers: Theoretical reflections on assessment for learning. *Journal of English For Academic, 4*(1), 40. https://doi.org/10.25299/jshmic.2017.vol4(1).505

Sin, K., & Muthu, L. (2015). Application of big data in education data mining and learning analytics: A literature review. *ICTACT Journal on Soft Computing: Special Issue on Soft Computing Models for Big Data, 5*(4), 1035–1049. https://doi.org/10.21917/ijsc.2015.0145

Southern Association of Colleges and Schools Commission on Colleges [SACSCOC] (2012). *Resource manual for the principles of accreditation: Foundations for quality enhancement.* https://sacs.fsu.edu/wp-content/uploads/2018/06/sacs-resource-manual-2012.pdf

Southern Association of Colleges (SACs) and Schools Commission on Colleges. (2005). *Resource manual for the principles of accreditation*. http://www.sgc.edu/president/departments/IE/files/SACSPrinciplesofInstitutionalEffectiveness.pdf

Stiggins, R. J., Frisbie, D. A., & Griswold, P. A. (1989). Inside high school grading practices: Building a research agenda. *Educational Measurement: Issues and Practice, 8*(2), 5–14.

Taras, M. (2005). Assessment – summative and formative–some theoretical reflections. *British Journal of Educational Studies, 53*(4), 466–478. https://doi.org/10.1111/j.1467-8527.2005.00307

Wells, C. S., & Wollack, J. A. (2003). *An instructor's guide to understanding test reliability.* Madison: University of Wisconsin, Testing & Evaluation Services. Retrieved from https://testing.wisc.edu/Reliability.pdf

INDEX

ABOUT THE AUTHOR

Melanie A. DiLoreto earned her Ph.D. in educational research, evaluation, statistics, and assessment from the University of Southern Mississippi. She holds a master's degree in educational leadership/school curriculum development and a bachelor's degree in elementary education. Dr. DiLoreto, a faculty member of teacher education and educational leadership at The University of West Florida, teaches courses in educational assessment, testing and measurement, program evaluation, research design, applied statistics, curriculum development, and advanced data analyses. Her research interests include various methodological approaches to investigating facets of educational assessment and student engagement in multiple learning environments.

www.ingramcontent.com/pod-product-compliance
Lightning Source LLC
Chambersburg PA
CBHW081125180926
51611CB00004B/53
* 9 7 8 1 5 1 6 5 4 1 7 0 6 *